DATE DUE

DEMCO 38-297

DEMCO

SOLAR ENERGY HANDBOOK

CHILTON BOOK COMPANY / RADNOR, PENNSYLVANIA

SOLAR ENERGY HANDBOOK

Theory and Applications

Power
Systems
Group/
AMETEK
Inc.

Copyright © 1979 by AMETEK, Inc.
All Rights Reserved
Published in Radnor, Pennsylvania, by Chilton Book Company
and simultaneously in Don Mills, Ontario, Canada,
by Thomas Nelson & Sons, Ltd.
Manufactured in the United States of America
Designed by Arlene Putterman

Library of Congress Catalog Card No.: 78-14646
ISBN: 0-8019-6776-7

2 3 4 5 6 7 8 9 0   8 7 6 5 4 3 2 1 0 9

CONTENTS

v

NOTE ON FORMULAS AND EQUATIONS

The formulas and equations in this handbook have been organized in a manner that is convenient for use with handheld programmable calculators. The primary computational features of each of the chapters may be incorporated by the reader into a magnetic card program for the Texas Instruments, Hewlett-Packard, or equivalent programmable calculators. This may be of particular value where numerous repetitive calculations occur in the worksheets for determining collector area in Chapter 7.

NOTE ON UNITS AND NOTATION

Both English and metric (SI) units are used in this handbook. English units appear in places where engineering or practical formulas result in order to conform to prevailing construction practices in the United States. Metric units are used in the theoretical sections and the unit notation is indicated for each formula and equation.

PREFACE

The objective of the *Solar Energy Handbook* is to assist in a greater understanding of solar energy utilization and solar energy products. It incorporates theoretical knowledge relevant to solar energy which has been generated through research by many investigators in diverse scientific fields over a great number of years. It also includes contemporary information gained through research and practical experience specific to the field of solar energy and its utilization. This handbook is not intended to be a comprehensive source to all aspects of solar energy; indeed, in a discipline which is evolving so rapidly, no single volume could be. But it can make contributions as a primer to theory and as a guide to practical applications.

The *Solar Energy Handbook* is the result of a cooperative effort of many individuals within AMETEK, Inc., including Phillips V. Bradford, John C. Bowen, Thomas R. Conlin, Robert A. Russell, and the AMETEK Power Systems Group.

SOLAR ENERGY HANDBOOK

CHAPTER 1
SOLAR ENERGY MARKETS

The domestic demand for energy now substantially exceeds the domestic supply and the trend is ominous. Figure 1–1 compares the estimated total demand to the composite of all of the traditional major energy sources—oil, natural gas, coal, uranium, and hydroelectric power—for the years from 1960 to 2000. The disparity between demand and supply is obvious, but there has always been a disparity since the first proto-man had to leave his shelter to search for wood. While the gap is not new, its increasing magnitude, as energy demand far outstrips supply, is alarming. We might still be able to accept this widening gap as a natural consequence of modern industrialized life so long as we could continue to increase supply. But a second glance at Figure 1–1 will show that our modest success in maintaining the supply curve in a degree of proximity to demand has not been a result of proportional increases in contributions from all energy sources. Of the five, hydroelectric power, coal, and natural gas have remained almost static while full development of the fourth, uranium-based energy, lies in the future. It is oil which has been almost wholly responsible for supply increases during the past decade and an increasing portion of that oil has been imported. Our need for foreign oil has become a dependence; without some remedial action begun today, that dependence can do nothing other than worsen.

TRADITIONAL RESOURCES

The employment of oil from foreign sources as a major component of our energy supply is totally unsatisfactory. It limits our freedom with the burdens of the international *quid pro quo* and strains our economy as the balance of payments is afflicted with increasing deficits. Moreover, even without unilateral action of the sort which the OPEC consortium initiated in 1973, the world competition for remaining reserves of petroleum will have the effect of a worsening price disadvantage and can cause conflict between otherwise peaceful nations which must compete in the same market for what has become the lifeblood of their industries and commerce. We will never have an overabundance of petroleum-based energy nor achieve complete independence from oil as an energy resource, but we must strive to narrow the demand/supply gap. We would be well advised to do it not with more imported oil and gas, but with equivalent domestic substitutes.

The obvious choice is to expand the supply of other traditional energy resources from domestic reserves, but the outlook is not bright. Domestic petroleum reserves are now declining as wellhead production exceeds the rate of new discoveries. There are new sources ahead, potential offshore reserves

FIGURE 1–1 The energy gap now filled by imported fuels will be closed with alternative energy sources, such as solar energy.

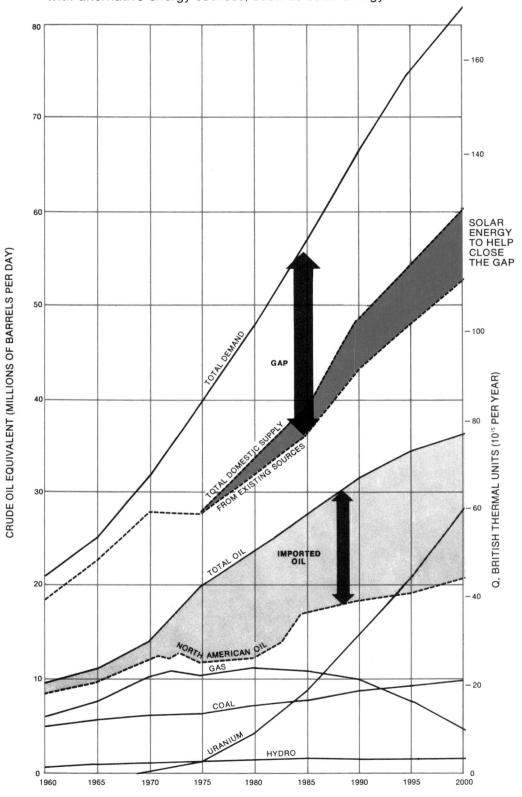

for example, but these will provide only partial and temporary relief. Natural gas is already in short supply, indeed has been rationed in some parts of the United States, and the prospects for new domestic supplies are less encouraging than those for oil. Even granting the unexpected discovery of oil and gas reserves sufficient to diminish the need for foreign oil, these resources are inherently finite. We cannot produce more than the eons have left for us and we are already well advanced in the depletion of our economically recoverable reserves.

Coal is our most abundant energy resource; in fact, a large part of the total world reserves are within the United States, but the use of coal presents many environmental risks, in addition to the safety hazards of mining. We might proceed with the utilization of coal wherever possible and ignore the damages of strip mining, the dangers of deep mining, and the pollution of an energy process based upon combustion. But coal is a less than ideal source. It must be mined, processed, loaded, transported, and unloaded before being put to use. It is bulky and unsuited to modern pollution-control requirements. And it will become more expensive, especially in the presence of greater demand, as the costs of its production rise.

Nuclear energy production utilizing radioactive fuels can be expected to increase its contribution as experience and wider use make it more cost effective. But there are problems here as well; the questions of nuclear waste disposal and nuclear safety engineering are of such gravity that a prolonged period of research and testing must precede wider use. Most importantly, nuclear energy, like modern coal-based energy, is fundamentally limited to large-scale central production. To utilize these two resources at a rate expansive enough to rapidly narrow the demand/supply gap will require an acceleration of favorable regulatory decisions. Made in haste, such decisions could result in potential environmental consequences which could be avoided by supplementing these traditional energy sources with a workable alternative.

ALTERNATIVE ENERGY SOURCES

The search for alternative sources of energy, supported by federal and private funds, has yielded a variety of possibilities with differing degrees of promise. They include coal gasification, coal liquefication, shale oil extraction, nuclear breeder reactors, nuclear fusion, and various forms of solar energy. Some of these possibilities have allied drawbacks similar to those of existing sources and almost all are in such a preliminary stage of development that their use in practical applications must await future technological advances. Only solar energy can be accurately said to have the advanced technology necessary to yield present time use and hold the potential of foreseeable, significant energy contributions in the future.

THE UNIQUE RESOURCE

Solar energy possesses characteristics which make it highly attractive as a primary energy source. It is based upon a continuously renewable resource which cannot be depleted and which is not subject to political control. It is available locally and does not require transportation. Of all energy sources, it is the least encumbered by environmental and safety hazards. And, most significantly, it is possible to collect, convert, and store solar energy with present technology. Future advances can be expected to improve performance characteristics and, perhaps, to reduce the cost of solar energy equipment.

There are unique problems, to be sure, which can affect the rate of growth. Energy can only be collected during daylight hours and then most efficiently on sunny days. Thus, some provision must be made for storage to accommodate nighttime and cloudy-day demands. The energy in sunlight is distributed through space in such a way that the amount collected, and therefore the energy yield of the system, is dependent upon the area intercepted. In addition, the sun is uncooperative enough to persist in its apparent motion in daily and seasonal ranges of considerable expanse across the sky with a consequent variation in the amount of power available. But all these disadvantages are outweighed by the twin benefits of the universality of the source and the long-term economy of operating a solar energy system.

Long-term operating economy, discussed in detail in Chapter 8, is another attractive characteristic of solar energy systems. Unlike traditional energy sources, however, the initial costs of tapping and converting the resource of the sun are relatively large per unit of power produced. Furthermore, the costs of those systems which are built to serve a single house or building must be borne by individual users who often lack the financial capabilities of large utilities and other central power generating authorities. For this reason, governmental in-

TABLE 1–1 Expected Energy Savings from Solar Technologies under the National Energy Plan*

(Fuel Displaced in Quads/Year†)

Solar Technology	1985	2000	2020
Hot water and space heating of buildings (direct thermal)	0.15	1.6	3–5‡
Process heat (direct thermal)	0.02	2.0	13
Wind energy conversion systems Electric utility	—	1.7	6.6
Solar thermal Electric utility	—	0.3	2.9
Photovoltaics Electric utility	—	—	0.2
Ocean thermal energy conversion systems	—	—	2.4
Biomass: electricity Wood	—	—	0.6
Biomass: fuels Wood	0.03	0.4	4.4
Total	0.2	6.0	33.1–33

SOURCE: MITRE Corp.—Less than 0.01 Quads.
* Although these estimates may be reduced due to delays in implementing the National Energy Plan, or increased as a result of federal policy and program alternatives now under consideration, they represent a consistent base upon which to evaluate changes.
† Electricity is converted to equivalent quads of primary energy based on 10,000 Btu/kWh.
‡ An estimate based on an extrapolation of the simulations up to the year 2000, adjusted with the likely addition of penetration of the gas and oil market.

centives for the development of solar energy and its financing at the local level may play a significant role in the market's development.

MARKET PROJECTIONS FOR SOLAR ENERGY

Long-term estimates of the contribution of solar power to the overall rate of energy supply in the United States have been made by the MITRE Corp., under contract to the Department of Energy (DOE). Table 1–1 shows these estimates, which include breakdowns for both direct and indirect solar energy applications. The power unit used in these long-range forecasts is the Q. One Q (or Quad) is 10^{15} (or one quadrillion) British Thermal Units (BTU) per year where one BTU is the energy required to raise the temperature of one pound of water by one degree Fahrenheit. Some appreciation of the magnitude of these future contributions may be gained by the estimation that one Q would raise the temperature of Lake Erie to the boiling point in one year.

SOLAR ENERGY APPLICATIONS

An attempt to list all present and potential solar energy applications would probably be forever incomplete since the breadth of present technology and the versatility of the process allow a wide range of innovation. There are three primary applications, however, already in use, which can help to delineate the size of specific markets. They are solar energy for water heating, space heating, and space cooling.

The Solar Energy Industries Association (SEIA) has projected these markets in terms of the square footage of collector area installed since collector panels are common to every system. These projections are summarized in Tables 1–2 through 1–5. These are important markets for two reasons. First, the applications involved now account for 20% of total energy consumption in the United States. Secondly, the principal fuels now used in these applications, including that portion which is first converted into electrical energy, are the two particularly scarce resources: oil and natural gas. Solar energy is expected to capture a substantial share of these markets within 10 to 15 years and will replace 1,000,000 barrels of oil (or its equivalent) per day. This can represent 5% to 10% of our continuing energy imports (primarily foreign oil), even after credit is taken for planned acceleration of other alternatives and for the effect of increased conservation efforts.

WATER HEATING

Solar water heaters are expected to become a major market factor for several reasons. Hot water requirements are year round, so the user's investment in a solar water heater will yield the maximum return. The necessary equipment is the least complex and smallest in size, thereby lending itself to retrofitting in existing installations. Furthermore, during the early years of solar market development, while the total number of solar installations is still modest

TABLE 1–2 Residential Solar Installations (Thousands of Square Feet of Collector Area): SEIA Forecast, February, 1978

	Water Heaters Only	Space Heating and Water Heating	Heating, Hot Water and Air Conditioning	Total
1977	3,800	1,060	40	4,900
1978	7,200	2,220	80	9,500
1979	12,000	3,440	160	15,600
1980	19,500	5,380	320	25,200
1981	27,600	9,250	650	37,500
1982	35,900	13,800	1,300	51,000
1983	42,500	19,000	2,500	64,000
1984	49,600	28,400	3,000	81,000
1985	58,000	43,000	5,000	106,000
1986	66,000	62,000	8,000	136,000
1987	71,500	111,000	12,500	195,000
1992	84,000	714,000	62,000	860,000

and long-term operating experience at a minimum, the smaller investment necessary for a solar water heater as compared to a solar space-heating unit will lead many users to limit their risk by keeping expense at a minimum.

Some impetus may be extended to the market by federal incentive tax legislation. The National Energy Act (NEA), passed in 1978, includes federal income tax credits for individuals installing solar energy systems. These federal tax credits, in conjunction with many state and local tax incentive programs, are providing a financial boost to potential customers. Forecasts indicated in Tables 1–2 through 1–5 include those incentives provided in the NEA but do not account for the market lag in 1977 and 1978 due to the delay in the final adoption of the NEA from its initial introduction in 1977. Table 1–2 shows the rapid growth of water heaters in the residential sector.

Table 1–3 includes SEIA estimates of the non-residential market for solar water heaters. Initial penetration of this market, including installations for

TABLE 1–3 Non-Residential Solar Installations (Thousands of Square Feet of Collector Area): SEIA Forecast, February, 1978

	Water Heaters Only	Space Heating and Water Heating	Heating, Hot Water and Air Conditioning	Process Heat	Total
1977	60	20	20	0	100
1978	150	50	50	50	300
1979	350	100	150	100	700
1980	700	200	300	200	1,400
1981	1,300	500	600	400	2,800
1982	2,300	900	1,100	800	5,100
1983	5,000	1,600	2,000	1,600	9,200
1984	6,800	2,700	3,500	3,200	16,200
1985	11,400	4,600	6,000	6,200	28,200
1986	18,500	7,500	10,000	11,000	47,000
1987	28,000	12,000	16,000	20,000	76,000
1992	100,000	70,000	100,000	320,000	590,000

TABLE 1–4 Total Installations in Service (Thousands of Square Feet of Collector Area), Residential and Non-Residential: SEIA Forecast, February, 1978

	Water Heaters Only	Space Heating and Water Heating	Heating, Hot Water and Air Conditioning	Process Heat	Total
1977	3,860	1,080	60	0	5,000
1978	11,210	3,350	190	50	14,800
1979	23,560	6,890	500	150	31,100
1980	43,760	12,470	1,120	350	57,700
1981	72,660	22,220	2,370	750	98,000
1982	110,860	36,920	4,770	1,550	154,100
1983	157,360	57,520	9,270	3,150	227,300
1984	213,760	88,620	15,770	6,350	324,500
1985	283,160	136,220	26,770	12,550	458,700
1986	367,660	205,720	44,770	23,550	641,700
1987	467,160	328,720	73,270	43,550	912,700

schools, hospitals, office buildings, laundries, and so on, is expected to be slower than in the case of residential heaters. This is a function of the diverse nature of the market and the fact that many of the individual applications are highly specialized. The ultimate size of the market will not be diminished, however, since a number of influences will promote greater acceptance. The inevitability of the end of traditional fuel sources has already been noted; another factor is the effect of regulatory philosophy. Experience has shown that, during periods of fuel shortage, the energy consumption of the commercial sector is more subject to curtailment or allocation than is residential consumption. Substitution of the solar alternative in industrial and commercial establishments could well become vital to uninterrupted operations.

SPACE HEATING

The logical second stage of market development is through the extension of basic solar water-heating systems to provide for space heating. Indeed, without

TABLE 1–5 Solar Dollar Volume, Including Inflation at 5% per Year (Millions of Dollars): SEIA Forecast, February, 1978

	Residential	Non-Residential	Total
1977	$ 92	$ 2.4	$ 94.4
1978	180	6.8	186.5
1979	300	15.5	315.5
1980	500	31	531
1981	750	63	813
1982	1,000	113	1,113
1983	1,300	207	1,507
1984	1,700	375	2,075
1985	2,200	660	2,860
1986	2,800	1,100	3,900
1987	4,100	1,800	5,900
1992	19,000	15,000	34,000

consideration of the associated minimum costs which make the simpler water-heating systems initially more attractive, it would seem a wiser choice to start with space heating, since the vast majority of these systems generate hot water which could supply residential requirements. In any case, space heating will widen the solar market since such systems are especially suited to new residential construction.

Non-residential space-heating markets, like those of non-residential water heating, may proceed at a somewhat slower pace. This is a consequence of the need for greatly increased capacity in each such installation with an accompanying increase in costs. It is a reasonable assumption, though, that as the supply pressures of traditional energy materials increase and further research yields a unit cost decrease in solar equipment, solar energy applications in non-residential use will achieve success. Tables 1–2 and 1–3 include market projections for both space-heating markets.

AIR CONDITIONING

Most practical methods of air conditioning require high-temperature sources of heat energy for efficient operation. This means that the collectors in a solar-powered system must be of higher efficiency than those required for space or water heating. Thus, though technically suitable solar air-conditioning equipment is available, it is cost effective in a very limited range of application. Even so, the rising costs of conventional energy sources will progressively widen the solar air-cooling market.

This market presents a contrast to the water and space-heating markets in that non-residential air-conditioning applications are expected to grow more rapidly than residential ones. The reasons are inherent in the characteristics of the demand load and the configurations of the equipment, new and existing. The cooling season for commercial buildings, stores, offices, and the like is longer than that for private dwellings; in the southern United States, it is very nearly a year-round demand. Solar air-conditioning equipment will, therefore, be better utilized and more cost effective. In addition, the configurations of equipment in existing conventional systems make the adaption to solar power doubly attractive. Since the majority of non-residential systems yield chilled water as their product, they can be retrofitted with a solar unit in which the collectors and solar heat-driven chiller supplement the existing equipment. Those systems already incorporating a heat-actuated chiller, and the number is substantial, need only add a solar collection system. Finally, the typical solar-driven, heat-actuated chiller requires the use of a cooling tower. Cooling towers are now widely used for heat rejection in commercial air-conditioning installations but are not generally used in residences. The operating complexity imposed by the cooling tower is therefore not a material deterrent to the use of solar air-conditioning in non-residential operations.

PROCESS HEAT

While water heating, space heating and air conditioning are expected to provide the highest initial demand for solar equipment, the direct thermal energy requirements of many process industries make them well suited for solar technologies. Table 1–3 demonstrates that before the year 2000 process heat installations are expected to surpass the balance of non-residential installations and, after residential space heating, account for the second highest amount of solar equipment demand.

Table 1–2 and 1–3 indicate the annual square feet of solar installations, while Table 1–4 shows the total amount of solar installations for each application area, based on a starting year of 1977. Both residential and non-residential sectors are combined. The table assumes proper maintenance and operation of all installed systems, with no systems retired during the ten-year forecast period.

Gross calculations of annual power produced in each type of system can be calculated by multiplying the total square feet of installations in service by appropriate conversion factors. International Business Services, in their *Forecast of Solar Energy Applications in the United States, 1977–1992*, presented the conversion factors used here. For water heating only, multiply the total square feet of collector in service by 0.00021 Quads (Q) per thousand square feet of collector. This assumes an overall national average of 210,000 BTU per year per square foot of collection area. Similarly, 0.000128 Quads per thousand square feet for combined space heating and water heating; 0.000204 Quads per year per thousand square feet for combined air conditioning, space heating and water heating; and 0.000145 Quads per year per square foot for process heat. The annual power calculated is in terms of actual BTUs produced and should not be confused with "Fuel Displaced" shown in Table 1–1. The latter takes into account the conversion efficiencies from the primary source of energy to the actual power available for end use.

The annual solar dollar volume, shown in Table 1–5, indicates an initially high rate of growth through 1982, after which a growth rate of approximately 30% is maintained for the next five years. From 1987 to 1992, the increased demand for residential space heating applications is expected to slightly increase the industry's total growth rate.

MARKET EXPANSION

MARKET PROJECTION VARIATIONS

Projections of solar energy markets made by other organizations have differed with the SEIA figures on either side; some studies conclude that the market will expand at a greater rate than that envisioned by SEIA, others project a more conservative market growth. The METREK Division of the MITRE Corporation, for example, has analyzed solar market development using two sets of assumptions, one based on a National Energy Plan (NEP) and a second based on an extension of recent trends, the Recent Trend Scenario (RTS).

Under NEP, a governmental program of incentives, disincentives, inducements, and prohibitions is assumed in-place to transfer demand from oil and natural gas to coal and nuclear, as well as solar and other inexhaustible resources. Under the RTS, projections are based on the escalation of oil and gas prices at relatively high rates, especially after 1990, as supplies of these fuels become more limited. Both scenarios emphasize the development and use of technological advances as soon as possible, and the application of cost-reduction curves in manufacturing and installation which has been the experience with many developing technologies in the past.

MITRE's projections of annual sales under each of the two assumptions illustrate one range over which these necessarily speculative projections may vary. Under NEP, annual sales in millions of square feet are projected to be 83.5 in 1985, 161.9 in 2000. The more conservative RTS assumptions yielded annual sales of 52.4 in 1985, 112.6 in the year 2000.

MONITORING THE GROWTH OF SOLAR ENERGY UTILIZATION

An important tool in monitoring the growth of the solar energy industry and one which establishes some credibility for future projections is supplied by the DOE. It is a summary report of solar collector manufacturing activity as measured by the total square footage of collectors produced. This report, which is issued every six months and made available by the National Energy Information Center, also includes a discussion of trends that become apparent in the manufacturing statistics.

As shown in Table 1–6, the results indicate that during the first half of 1978 manufacturing activity in medium and high-temperature solar collectors increased 17% to 2,681,244 square feet over the second half of 1977. This increase is low when compared to the 22% increase from the first half of 1977 to the second half of 1978 and the 117% from 1976 to 1977.

The DOE statistics also include data on the manufacture of low-temperature collectors, which are used primarily for heating swimming pools and greenhouses. The annual production rate of these collectors is approximately 3,000,000 square feet, a total that has been declining since a peak production rate of 3,875,808 square feet in 1976.

BASIC ASSUMPTIONS

The market forecasts of this chapter are based on certain conservative assumptions concerning the future. It is assumed that the rate of inflation will be 5% per year, that electricity utility rates will rise at a 7% annual rate, and that fuel oil and natural gas prices will rise by 10% each year. It is also assumed that the unit cost of a solar installation will gradually decline as the cumulative experience grows.

None of these assumptions can be made with certainty and variations in the assumed quantities can affect the rate of growth for solar equipment in either direction. For example, if the rate of inflation drops below 5%, it should become easier for customers to borrow for investment in solar energy systems. However, the inflation rate may be, at least in part, influenced by the domestic availability of energy. Similarly, the manufacturing cost of a solar system may rise with the cost of energy and with inflation, despite the economies of higher volume production.

TABLE 1-6 Solar Collector Manufacturing Activity: 1975–1978

Solar Collector Annual Production Rate (square feet)	Total 1975		Total 1976		Jan.–June, 1977		July–Dec., 1977		Jan.–June, 1978	
	Manu-facturers	Square Feet	Manu-facturers	Square Feet	Manu-facturers	Square Feet	Manu-facturers	Square Feet	Manu-facturers	Square Feet
Medium Temperature and Special Collectors										
greater than 50,000		*	9	692,204	18	861,152	7	568,190	9	704,567
10,000–50,000		*	38	829,465	63	840,688	47	1,065,808	60	1,310,133
greater than 10,000	19	524,873	(47)†	(1,521,669)†	(81)†	(1,701,840)†	(54)†	(1,633,998)†	(69)†	(2,014,700)†
2,000–10,000	35	149,097	76	333,752	60	161,088	123	570,687	116	571,382
1,000–1,999	14	19,969	33	44,615	21	15,047	51	62,451	54	69,725
less than 1,000	50	23,321	47	24,694	24	6,865	65	24,210	62	25,437
Total	118	717,260	203	1,924,730	186	1,884,840	293	2,291,346	301	2,631,244
Low Temperature‡	13	3,025,956	19	3,875,808	15	3,222,208	49	2,913,243	56	3,594,630

* Not shown separately before 1976.
† Number total of two preceding numbers in same column but not added for final total.
‡ Low temperature collectors are shown separately because they are generally made of plastic or rubber and are used almost exclusively for applications below 100°F.

11

Efforts to delineate the specifics of future reality in solar energy markets is, like all forecasting, an inexact art. The considerations noted above, as well as others, can have effects greater or lesser than present speculation may assign to them, and each of those effects will be, in turn, magnified or diminished by other influences impossible to foretell. But there is one final consideration that should be noted because it could very well reduce the others to insignificance. It is the simple matter of scientific acceleration.

The appeal of solar energy has tantalized man for thousands of years. The Greek physician and writer Galen recorded the use of "burning mirrors" situated on the harbor walls which Archimedes used to set fire to the Roman fleet attempting to invade Syracuse in 212 B.C. Serious experimentation in solar energy has been pursued for almost as long. But, as the later sections of this handbook will make apparent, much of the basic technology and most of the practical knowledge which make solar power usable today have been developed during the past century. It is reasonable to anticipate that, as research accelerates, the coming decades will produce a large number of technical advances, perhaps a major breakthrough, in the implementation and manufacturing processes used in solar energy. With these advances the costs could be greatly reduced, and the efficiency and utility of solar systems greatly enhanced, with an equivalent spur to market development. Such quantum leaps have occurred in other fields, space exploration and medicine are but two examples, and to great benefit. We should not formulate plans based solely on this possibility, but there is no reason not to recognize its potential.

CHAPTER 2
SOLAR PHYSICS AND
GEOMETRY

Science is a fragmented and accumulative process. Its practitioners often work independently, theorizing, experimenting, proving, but pile the resultant conclusions in a vast common storehouse. Discoveries are sometimes triumphs, but triumphs of the moment, for each unveiled fact quickly assumes the anonymity of data, history to be drawn from the common fund when needed, by discoverer or stranger alike, as support for future theories and further questioning. And, while each discipline in science is separate and distinct, a community concerned in the main with its own factual past and theoretical future, it is not independent. It is benefited by, and often depends upon, the work performed in other disciplines; physics, for example, reaches into chemistry for guidance and information while moving toward new knowledge.

This intramural exchange as the basis for progress is most evident in today's practicality of solar power. The debts are many, for the achievement of solar energy utility has drawn on the knowledge gained in physics, chemistry, thermodynamics, and a hundred others; similarly, its application equipment is an amalgam of metallurgy, semiconductor technology, fluid mechanics and another legion of contributors. Because of this, a complete study of solar energy entails the complexities of optics, structural engineering, heat exchange, electrical engineering, and many other disciplines, treatments which are far beyond the scope of this handbook since the object here is practical knowledge useful in today's applications. It will be valuable, however, when considering specific applications in later chapters, to have an understanding of a few basic theoretical concepts which, since they are fundamental, are concerned with that most fundamental part of all solar energy study: the sun.

The sun, like all functioning stars, is a gaseous sphere in a state of such complete and violent agitation that "turbulent" seems a pale description. Columns of gases, heated by core temperatures of many millions of degrees, fume up to the surface, lose some of their heat, and descend to be heated again. Giant magnetic storms, sunspots, split the seething cauldron of the solar gases to add their own definition of bedlam. And, as if to extend the violence past imagination, immense solar flares, cataclysmic eruptions of the surface, occur periodically to spew a bewildering array of radiation outward into space—radio waves, gamma waves, physical fluxes, visible light, ultraviolet light, infrared light, and x rays—energy enough in a single solar flare to satisfy earth's energy needs for 100,000 years. All of it is raw, naked power, the prime physical force of the universe, and all of it originates within the tiny simplicity of the hydrogen atom.

For 5 billion years the sun has gathered the elements from the vast reaches of interstellar space by gravitational attraction until, today, 1% of its mass is composed of traces of all elements, 24% is helium, and 74% is hydrogen. This is a necessary predominance, for hydrogen is the fuel for the sun's thermonuclear furnace. The energy-producing process has been duplicated on earth only in the hydrogen bomb where hydrogen transmuted into helium and the higher elements leaves a surplus of energy derived from the intranuclear "glue" which binds the particles of all nuclei together. The effects of this transformation are gargantuan, but the process itself and the vessel which contains it, the sun, are stable. When viewed on the mammoth scale of solar time and place, it must end at some point either in a white dwarf star shriveling in the absence of hydrogen fuel or in a supernova of incomprehensible explosive force. But, when framed in the lilliputian scale of human interests, the sun can be safely regarded as in eternal existence.

The sun's stability is a product of transference phenomena within the solar sphere. Energy is conveyed to the surface, the photosphere, where it is released into space to balance the stream erupting from the interior furnace. The photosphere consists of a plasma of ionized gases which dispose of energy in several ways. The solar wind, cosmic rays, and radiated neutrinos expel some of the vast output, but the majority of radiated energy is in the form of electromagnetic waves characteristic of the plasma's elemental composition. These waves are combined into the phenomena we know as heat and sunlight.

THE SOLAR SPECTRUM

The sun's electromagnetic radiation occurs in a large number of pure waves of different wavelengths combined into the "extra-terrestrial solar radiation" depicted in Figure 2–1. Also shown, in order of increasing magnitude of wavelength, are its components: x rays, ultraviolet light, visible light, infrared light, radiant heat, and radio waves. Most of the sun's energy which can be recovered in a useful way is in the form of ultraviolet, visible, and infrared light.

All pure electromagnetic waves are characterized by wavelength, λ, and frequency, f, through the relationship:

$$\lambda f = c \tag{2–1}$$

where c is the speed of light in a vacuum: 2.998×10^8 meters/second.

The unit of energy intensity used in Figure 2–1 is the Langley, named for Samuel Pierpont Langley (1834–1906), an American scientist and aeronautical pioneer who made major contributions to methods of measuring solar energy. One Langley (Ly) is one calorie of energy incident normal to one square centimeter of area; one Langley per minute (Ly min^{-1}) is equivalent to 221.2 BTU per hour per square foot. The spectral energy, S, in Figure 2–1 is expressed in Langleys per minute per micron (μ) of wavelength on a logarithmic scale. The unit for wavelength, λ, is the micron, μ, or one millionth of a meter, also on a logarithmic scale. The logarithmic scales are used so that a wide range of conditions can be displayed.

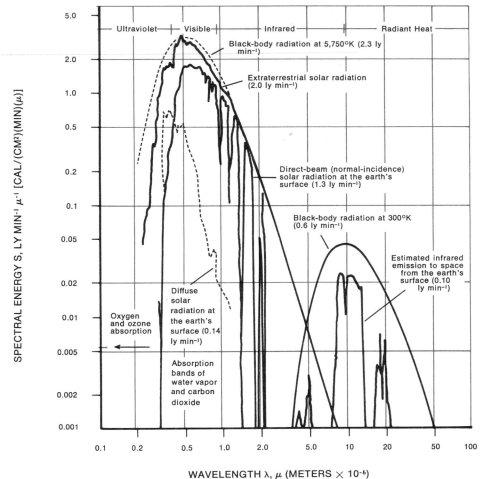

FIGURE 2–1 The solar spectrum.

In an ideal world one would seek to capture as much of this prodigious outpouring as possible, but the width of the solar spectrum in general and the longer radiant heat wavelengths in particular, are not ideal. In many cases, as will become apparent in the section on selective coatings, it is better to utilize a collector which absorbs sunlight and reflects the longer wavelengths. Similarly, collector covers should be constructed of materials which transmit radiation best within the solar spectrum while absorbing or reflecting radiant heat waves.

THE SOLAR CONSTANT

The solar constant is the reference point from which a determination can be made of the amount of solar energy that is available for utilization. It is defined as the total energy emitted by the sun per unit of area perpendicular to the sun's rays in near-earth space at the average earth distance from the sun per unit of time. As might be suspected, this does not represent power available to earthbound collectors, since the changing geometry of the sun as it moves

TABLE 2–1 The Solar Constant, I_0

$I_0 =$	1354	Watts per square meter
$=$	1.354	Kilowatts per square meter
$=$	429	BTU per hour per square foot
$=$	1.94	Langleys per minute = calories per square centimeter per minute
$=$	4870	Kilojoules per hour per square meter
$= 4.87 \times 10^6$		Joules per hour per square meter
$=$	1.52	Horsepower per square yard

across the sky and the interposition of atmosphere and climate all have an effect. In addition, though the solar constant represents a useful reference, its stated value differs slightly in various source publications. This is a consequence of the measurement techniques used, spacecraft instrumentation or terrestrial astronomical data, both of which contain sources of inaccuracy. Other variances in absolute value result from natural variations in the sun's intensity, sunspot population, and temporary spectral anomalies, but the greatest natural variation results from the elliptical configuration of the earth's orbit. In early January the earth occupies its nearest approach to the sun, thereby increasing the solar constant by about 3%. In early July the average value declines by about 3% when the earth reaches its greatest distance from the sun. Table 2–1 expresses the average value of the solar constant in different systems of units.

BLACKBODY RADIATION LAWS

As is true in many disciplines, the foundation for understanding real events is an understanding of an ideal case which exists only in theory. Economists, for example, use the impossibility of perfect competition as their laboratory case, then proceed to study the ways in which reality approximates or departs from that ideal.

The theoretical case underlying the reality of solar energy utilization is that of the blackbody. A blackbody, so called because of the resulting color, is defined as an object that absorbs all, and reflects none, of the radiation incident upon it. No perfect blackbody exists, but the concept is important because the radiation laws derived from that perfect case can then be used to yield a relationship between the properties of real objects in actual radiative environments. Reality here, of course, is the energy exchange between the originating sun and the receiving object on earth.

A detailed discussion of blackbody radiation laws and the theoretical derivation of relevant equations is contained in Appendix A. Though complete familiarity with this theoretical background is not indispensable to understanding solar energy utilization, the material does form the basis for practical application. As such, it is well worth study.

Real objects are characterized by their absorption and emission compared to those of a blackbody at the same temperature. The absorption and emission of an object are expressed as coefficients, α and ϵ respectively, ranging from zero (for no absorption and no emission) to one (for absorbing and emitting as well as a blackbody at the same temperature). Both the absorption and emission coefficient depend upon the temperature of the object, the material of which it's made, and surface characteristics such as roughness, coatings, corrosion, and features. In general, the absorption and emission of an object vary with temperature, wavelength, and angle of incidence, with the absorption and emission coefficients usually expressed as average values over a range of these variables.

When radiation is incident upon an object that does not completely absorb it, the portion that is not absorbed is either reflected by, or transmitted through, the object. The coefficient of transmittance, τ, represents the fraction of incident radiation that passes through a transparent or translucent object. Transmission can be direct (as through clear glass), diffused (frosted glass), or in multiple refracted beams (certain crystals). The coefficient of reflectance, ρ, represents the fraction of incident radiation that is reflected from an object. Reflectance can be diffuse (as from snow), or specular (as from a mirror). In general, transmittance, τ, and reflectance, ρ, vary with temperature, wavelength, angle of incidence, and polarization.

The coefficient of transmittance and reflectance, τ and ρ, and the ratio of α to ϵ (see equation A–8) are significant characteristics of any solar absorber. The α/ϵ ratio is often regarded as an important characteristic, since it relates to the temperature which the absorber can achieve if all heat losses other than radiative can be eliminated. It would be misleading, however, to assume that a high value of α/ϵ implies greater energy collection by the object, since the amount of energy collected increases with increases in α even if the ratio α/ϵ remains fixed.

APPARENT MOTION OF THE SUN

The relation between a particular site on this planet and the sun must be expressed with terms which include accommodation for the geographical location on the earth, the status of the earth's rotation (usually expressed as the time of day), and the relationship between the positions of the earth and sun in space. Each has its effect on the absolute value of solar energy available for utilization.

The first, geographical location, is the easiest with which to deal. Any position on the earth can be designated by its latitude, longitude, and elevation above sea level. The position of the sun as observed from this site is designated by the angle of elevation above the horizon and its azimuth (the angle formed by the sun's projection to the horizon and due south.)

The position of the earth during its annual journey around the sun can be expressed by the angle of declination, D, which represents the amount by which the earth's north polar axis is tilted toward the sun. A near approximation for D is:

$$D = 23.44 \sin (360 \, d/365) \text{ degrees} \qquad \textbf{(2–2)}$$

where: D is the angle of declination in degrees

d is the number of days following the vernal equinox which usually falls near March 21

and where the number 23.44 is the angle, in decimal degrees, between the earth's axis and the plane of the ecliptic (earth's orbital plane).

The time of day, for solar purposes, is expressed in terms of the hour angle, H, which represents one 24-hour day as 360 degrees of angle.

$$H = 15\,t \text{ degrees} \tag{2-3}$$

where: t is the time in hours (decimally) from solar noon

H is the hour angles in decimal degrees.

Each time period of 4 minutes advances the hour angle by one degree.

The local time, or clock time, at which solar noon occurs will differ from 12:00 noon as a result of several effects which are incorporated into the following equation:

$$t_{sn} = 12:00 + 4\,(Y - Y_0) + t_c, \text{ minutes}$$
$$+ \text{ 1 hour if daylight savings time is in effect} \tag{2-4}$$

where: Y is the longitude of the site in decimal degrees

Y_0 is the longitude of the center of the local time zone (see Table 2–2)

t_c is the correction from the equation of time (see Figure 2–5 or Equation 2–5)

t_{sn} is the time at which solar noon occurs.

Within a given time zone, local time is established as a discrete multiple of one hour from the time at the zero meridian at Greenwich, England; thus, a

FIGURE 2–2 The seasons of the year are caused by the tilt of the Earth's axis.

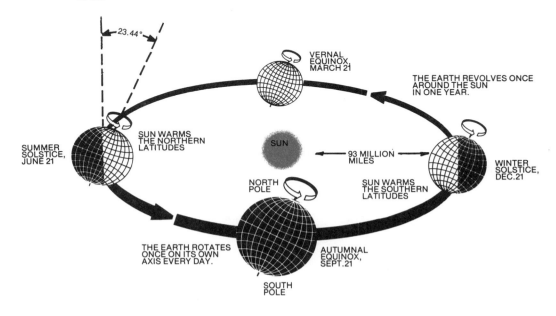

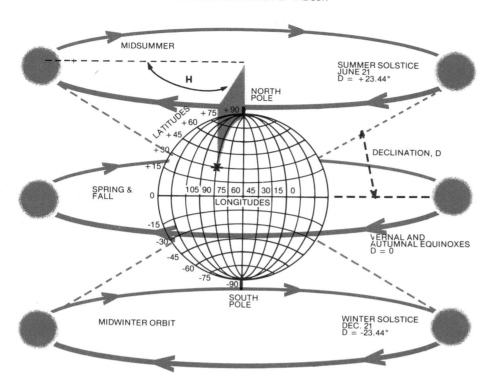

FIGURE 2–3 From an earth regarded as stationary, the apparent motion of the sun is described by the hour angle, *H,* which depicts the time of day, and the declination, *D,* which depicts the time of year. A site on the earth is specified by its latitude, *L,* and longitude, *Y.* The hour angle increases uniformly with 15 degrees corresponding to each hour, or one degree every four minutes, from solar noon.

location west of the zone's center will experience its solar noon at a later time than at the center and a location east of the center will experience an earlier solar noon. In specific terms, solar noon will occur four minutes later for each degree of longitude west of the center of a time zone.

The longitude of the centers of U.S. time zones and representative cities are shown in Table 2–2. This table provides the value of Y_0 for use in Equation

TABLE 2–2 Time Zones and Longitudes

Time Zone	Y_0, Longitude at Center	Representative Cities
Eastern	75°	Camden, New Jersey
Central	90°	Memphis, Tennessee
Mountain	105°	Denver, Colorado
Pacific	120°	South Lake Tahoe, California

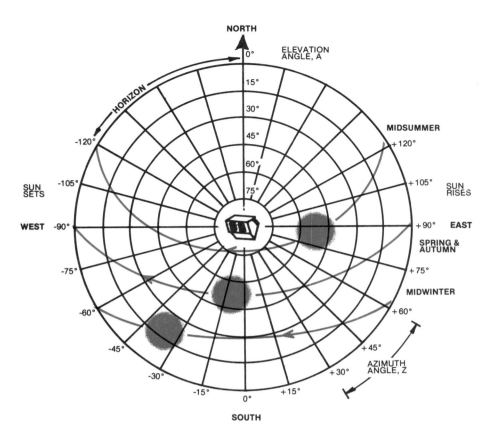

FIGURE 2–4 In a plan view of a solar site from above, the elevation and azimuth angles of the sun may be plotted on a normal projection of a celestial hemisphere. The site in the center is located at 40° north latitude.

2–4. If the longitude of the site is not known, it may be estimated by determining the distance in miles east or west of the center and multiplying by 69 times the cosine of the latitude of the site. This will yield the difference in longitude between the site and the representative city in degrees. Multiplication by four will yield the number of minutes that solar noon will follow 12:00 noon if the site is west of the representative city or the number of minutes by which solar noon will precede 12:00 noon if the site is east of the representative city.

Another correction, t_c, is needed to establish the local reckoning of solar time. It is due to two different effects of the earth's motion around the sun. The first is the cumulative advance and retardation of solar time with seasonal changes in declination; the other is the cumulative advance and retardation of solar time with variations in the earth's orbital speed resulting from its slightly elliptical orbit around the sun. These factors are both taken into account by the time correction, t_c, shown in Figure 2–5 and applied in Equation 2–4. The curve shown in Figure 2–5 is known as the "equation of time" and is represented graphically on many globes as a distorted figure-eight known as an analemma. The equation of time may also be represented to within an accuracy of one minute by the formula:

$$t_c = 9.9 \sin (2y) + 7.3 \sin (y - 13°) \text{ minutes} \qquad \textbf{(2–5)}$$

FIGURE 2–5 The equation of time.

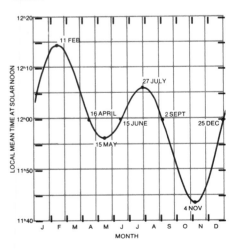

where $y = \dfrac{360}{365}(d + 80)$ is the "year angle" in degrees, and d is the day number defined following Equation 2–2.

The time of solar noon, t_{sn}, is the time during the day when the sun achieves its highest elevation angle above the horizon, and when its azimuth angle is zero. Thus, it is located in the sky at a direction of due south. One method of determining directions is to note the direction of the shadow of a plumb vertical at solar noon. It will be oriented in a north–south direction.

The local times at which sunrise and sunset occur can be found by adding and subtracting an hour angle, H_s, from that of solar noon and converting to time units. The hour angle, H_s, is determined by:

$$\cos H_s = -\tan D \tan L \qquad \text{(2–6)}$$

where: H_s is the hour angle between solar noon and sunset or sunrise

  D is the declination angle from Equation 2–2

  L is the latitude of the site.

The geocentric time of sunrise, t_{sr}, and the time of sunset, t_{ss}, are given by:

$$t_{sr} = t_{sn} - \frac{H_s}{15} \qquad \text{Hours (decimal)}$$

$$t_{ss} = t_{sn} + \frac{H_s}{15} \qquad \text{Hours (decimal)} \qquad \text{(2–7)}$$

where: t_{sn} is the time of solar noon given by Equation 2–4.

The actual time of sunrise and sunset are earlier and later, respectively, than that predicted by Equation 2–7 because of the refractory distortion of the atmosphere and slight parallactic advantage due to the elevation of an observer.

The astronomical definition of sunrise and sunset is based on the appearance and disappearance, respectively, of the upper limb (edge) of the sun as it moves across a level horizon. The limb of the sun is 16′ of arc from its center and atmospheric refraction accounts for about 34′ of arc in a vertical direction. The correction to be applied to Equation 2–7 to compute the astronomical sunrise and sunset (as distinct from the geocentric) requires an addition, ΔH_s, to the value of H_s, given by:

$$\Delta H_s = \frac{1}{0.3 \cos L \cos D \sin H_s} \qquad \text{(2–8)}$$

where ΔH_s will be expressed in degrees of angle.

Equations 2–2 through 2–8 can be programmed into card programmable calculators for convenient application.

The position of the sun can be determined approximately by the following equations for its elevation above the horizon, A, and its azimuth angle with respect to due south, Z. All angles are in degrees.

$$\sin A = \cos D \cos H \cos L + \sin D \sin L \qquad \text{(2–9)}$$

$$\sin Z = -\frac{\cos D \sin H}{\cos A} \qquad \text{(2–10)}$$

where: A is the angle of elevation of the sun, positive above the horizon and negative below the horizon

Z is the azimuthal angle of the sun, positive to the east and negative to the west if measured from the south

D is the declination angle from Equation 2–2

H is the hour angle from Equation 2–3, negative in the morning and positive in the afternoon

L is the latitude of the site.

Equation 2–10 represents an interesting geometrical fact. It states that the product $\sin Z \cos A$ is independent of the latitude from which the sun is observed.

The declination, D, is the elevation of the sun to an observer at the North Pole and the hour angle, H, is the negative value of its azimuth there. Equations 2–9 and 2–10 constitute a transformation of solar elevation and azimuth angles as reckoned from the North Pole to any desired location on the Earth.

In the application of Equation 2–10 for determining the azimuth angle of the sun, there is an important precaution to observe. That is when the azimuth angle, Z, exceeds 90°, the value of $\sin Z$ will be the same as for an angle less than 90° by an amount equal to that by which the solar azimuth exceeds 90°. Thus, once the value of $\sin Z$ is determined by Equation 2–10, the determination of Z is sometimes ambiguous. The ambiguity is resolved if the azimuth angle is redefined as being referred from the north if the sun is in the northern sky and from the south if the sun is in the southern sky.

Another method of resolving the ambiguity is to use the following equation to obtain the azimuth, Z:

$$\tan \frac{Z}{2} = \frac{\cos D \sin H}{\cos L \sin D - \sin L \cos D \cos H - \cos A} \tag{2-11}$$

This equation, when solved for Z, using the principal value of the inverse tangent function, provides the correct azimuth angle under all conditions. In the southern hemisphere, however, the azimuth is referred to south, just as in the northern hemisphere. The usual convention in the southern hemisphere is to refer azimuth angles to the north, which is automatically accomplished in the more simply expressed Equation 2–10.

In the southern hemisphere, the latitude, L, should be entered into equations 2–9 and 2–11 as a negative number. If Equation 2–10 is used, the azimuth should be referred to north when the sun is in the north and referred to south when the sun is in the south; however, the sign of Equation 2–10 should be reversed from negative (−) to positive (+). If, instead, Equation 2–11 is used to determine the sun's azimuth, Z, then it will be referred to south at all times when either northern latitudes or southern latitudes are used.

The accuracy of these formulas (Equations 2–9 through 2–11) is based upon the assumption that the earth is a sphere and that daily variation in declination, D, may be ignored. In fact, the earth is slightly pear shaped, the declination changes slightly every day, there is some parallax in the direction of the sun as seen from different points on the earth, the earth's motion is affected by the moon, and there are long-term changes in the obliquity of the equinox,

FIGURE 2-6 Solar geometry.

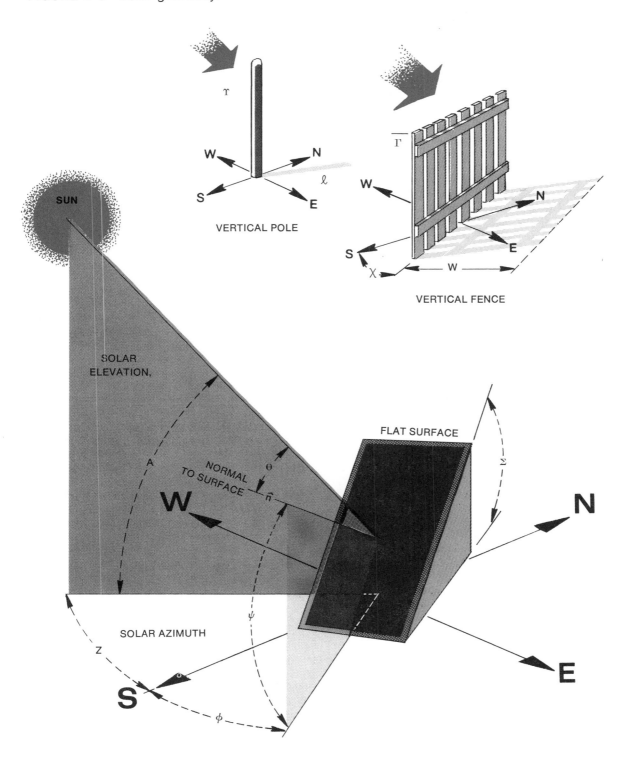

VERTICAL POLE

VERTICAL FENCE

SUN

SOLAR ELEVATION,

NORMAL TO SURFACE

FLAT SURFACE

SOLAR AZIMUTH

which itself precesses and nutates. There are, additionally, totally unexplained changes in the earth's speed of rotation that some astronomers relate to changes in its internal structure. Fortunately, all these effects are small and can safely be ignored in most solar energy applications. The formulas presented here can be considered accurate to within the angular diameter of the sun. The sun's average angular size is $16'0''$ of arc from its center to its edge, or slightly more than one-half degree from side to side as viewed from the earth.

A flat plane surface with normal (perpendicular to the surface) directed at an elevation angle, ψ, from the horizontal and at an azimuth angle, ϕ, from due south forms an angle, θ, with the sun. The angle, θ, is the angle between the sun's rays and the normal to the plane surface and is given by:

$$\cos \theta = \sin A \cos \Sigma + \cos A \sin \Sigma \cos (Z - \phi) \qquad \textbf{(2–12)}$$

where the geometry is shown in Figure 2–6.

The angle, Σ, is the tilt angle of the plane as measured from a horizontal surface. By definition, it is the complement to ψ, or $\Sigma = 90° - \psi$. The angle, ϕ, is the direction in which the plane is tilted measured from $\phi = 0$ at due south and increasing positive as it faces a more easterly direction, negative to the west. The angle, θ, between the sun's rays and the normal to a plane surface is known as the incident angle.

SOLAR INSOLATION

The incident angle, θ, is important because it determines the magnitude of solar insolation. Solar insolation is defined as the solar intensity that is incident normal upon a unit area of the plane surface. The direct beam of solar energy incident to a tilted plane surface is given by:

$$I_b = I_n \cos \theta \qquad \textbf{(2–13)}$$

where: I_b is the direct beam solar insolation on the tilted plane surface in BTU/hr-ft²

I_n is the solar insolation or direct normal intensity of the sun's radiation in BTU/hr-ft² (Equation 2–14)

θ is the incident angle given by equation 2–12.

The direct normal intensity of the sun's radiation, I_n, on a clear day is given by:

$$I_n = I_r e^{-mB} \text{ BTU/hr-ft}^2 \qquad \textbf{(2–14)}$$

where I_r and B are constants given in Table 2–3 for each month of the year and m is the air mass given by:

$$m = \frac{1}{\sin A} \qquad \textbf{(2–15)}$$

where A is the elevation angle of the sun given by Equation 2–9. The values of B in Table 2–3 are only considered accurate in latitudes from 24 to 56 degrees. Some error may occur if these values are used in equatorial or polar latitudes where seasonal changes and atmospheric conditions are different.

TABLE 2–3 Values of I_r and B for Equation 2–14

Month	I_r (BTU/hr-ft²)	B
Jan.	390	0.142
Feb.	385	0.144
Mar.	376	0.156
Apr.	360	0.180
May	350	0.196
June	345	0.205
July	344	0.207
Aug.	351	0.201
Sept.	365	0.177
Oct.	378	0.160
Nov.	387	0.149
Dec.	391	0.142

Although the values for I_r and B are determined empirically to fit measured radiation data, there is some basis for determining them through calculation. The values of I_r represent the effect of the upper atmosphere in removing a certain spectral portion of solar energy as the sun's rays pass through, and take into account the variation in distance between the earth and the sun. An approximate formula for I_r is given by:

$$I_r = 0.862 \, [I_0 + 27 \cos (y - 8°)] \text{ BTU/hr-ft}^2 \tag{2-16}$$

Where y is the year angle defined as following Equation 2–5 and $I_0 = 429$ BTU/hr-ft² is the solar constant from Table 2–1.

The values of B correspond to seasonal changes in the total air mass as a result of temperature, humidity, and biological debris which affect the transmission of solar energy through the atmosphere. An approximate formula for B is given by:

$$B = 0.172 - 0.033 \cos (y - 11°) \tag{2-17}$$

EFFECTS OF ELEVATION ABOVE SEA LEVEL

The availability of direct-beam solar energy tends to increase with the elevation of the site because of the reduced air mass through which the sun's rays must travel. At Denver, Colorado, which is about one mile above sea level, the direct normal solar insolation is greater than at Wilmington, Delaware, which is at the same latitude as Denver, but only a few feet above sea level.

An approximate method of determining the effect of altitude on solar insolation can be calculated by dividing the atmosphere into two layers, upper and lower. Certain portions of the sun's spectrum are quickly absorbed in the upper atmosphere, leaving an amount, I_r, given in Table 2–3 or Equation 2–16. Thus, the effect of elevation is not reflected in any change in I_r. The lower atmosphere, which extends to about 80,000 feet, has an effect on solar insolation that is in proportion to the air mass through which the sun's rays must travel. Since the density of air decreases nearly exponentially with altitude, an approximate formula for the air mass at elevation, E, is given by:

$$m' = me^{-E/E_0} \tag{2–18}$$

where: m' is the air mass at elevation E

m is the air mass at sea level given by Equation 2–15

E is the elevation above sea level of the site, in feet

E_o = 30,000 feet, is the approximate elevation where the atmospheric density is $1/e = 37\%$ of the standard atmospheric density at sea level.

To account for the effects of site elevation, m' should be substituted for m in Equation 2–14.

There is a limit to the accuracy that can be obtained in formulas for I_r, B, and m' because of the variability of the atmosphere. Each of these quantities depends on site elevation and latitude to some extent and also on specific features of a local region (in decreasing order of importance): pollution, moisture, pollen count, wind-blown dust, and volcanic dust.

The effect of site elevation is often treated by the use of an atmospheric clearness number which is to be multiplied by the direct normal solar intensity, I_n. The advantage of the clearness number is that it also accounts for the presence of high levels of moisture in low-elevation areas and compensates for the variations in the value of B due to changes in latitude. Table 2–4 summarizes the clearness numbers in the various non-industrial regions of the United States.

TABLE 2–4 Clearness Numbers Used as a Multiplier for I_n

Geographic Region	Clearness Number
Gulf Coast and Florida	Winter: 0.95 Summer: 0.90
Deep South	All year: 0.90
Middle South	All year: 0.95
Mid-Atlantic, New England, Midwest, Lower Great Lakes to Texas Panhandle	All year: 1.00
East Canadian Border, Upper Great Lakes, Northern Plains	All year: 1.05
Rocky Mountains	Winter: 1.05 Summer: 1.10
West and Southwest Desert Areas	Winter: 1.00 Summer: 1.10
West Coast	Winter: 0.95 Summer: 1.05

The determination of solar insolation at any location requires knowledge of climatic factors, as well as the geometric relationships between the earth and the sun. In this chapter, the geometric relationships were used to determine the amount of available direct-beam solar energy that can be realized on a surface of arbitrary orientation. Starting with the time of day, date, and location of a site, Equations 2–9 and 2–10 or Equations 2–9 and 2–11 provide the position of the sun in the sky. From this and the orientation of a plane surface, Equation 2–12 is used to determine the incident angle, θ. The incident angle is used to convert the direct normal solar insolation, I_n, to the value realized on the plane surface by Equation 2–13.

Direct normal solar insolation can be approximately determined by the methods shown in Equations 2–14 and 2–15, using empirical values for I_r, B, and clearness number.

Since no formula for clearness number is provided, readers may find that it will suffice to use an appropriately modified value for E, the site elevation above sea level, in Equation 2–18. In most areas, the actual altitude is reasonably accurate; however, in the Gulf coastal areas, Florida, and the deep and middle south, the effects of moisture and biological debris can have the effect of producing conditions comparable to an elevation somewhat below sea level. It may be more appropriate to rely on the clearness number in that region and in areas where the clearness number shows a marked seasonal variation.

There is no substitute for measured values of solar insolation, such as will be shown in Chapter 3. Measured values are usually based on a horizontal orientation where the incident angle is the 90° complement to the elevation angle of the sun, A. The relationship between the direct normal insolation, I_n, and the horizontal value of beam insolation is thus:

$$I_n = \frac{I_{Hb}}{\cos(90 - A)} = \frac{I_{Hb}}{\sin A} = m I_{Hb} \qquad \textbf{(2-19)}$$

where: I_{Hb} is the measured value of beam solar insolation on a horizontal surface

I_n is the direct normal solar insolation

A is the elevation angle of the sun given by Equation 2–9

m is the air mass given by Equation 2–15.

This method of determining the direct normal solar insolation, I_n, requires measured values of the beam solar insolation on a horizontal surface. Most solar insolation measurements on a horizontal surface, however, include diffuse and reflected radiation from the sky and clouds so that some accounting must be made for these effects. Diffuse radiation is treated in Chapter 3, since it is primarily dependent on climatic conditions.

SHADOW SIZES

The size of shadows cast by objects placed in the sun can be important in applications of solar engineering, since shadows cast by trees and buildings can interfere with the performance of solar collectors. For the same reason, linear arrays of solar collectors should be spaced so that their shadows do not interfere with the collection of adjacent rows during the times when optimum performance is desired.

A vertical pole of height, Y, casts a shadow on a flat horizontal surface with length l, given by:

$$l = Y/\tan A \qquad (2\text{–}20)$$

where: l is the length of the shadow of a vertical pole of height, Y, expressed in the same units as Y.

A is the elevation angle of the sun given by Equation 2–9.

A vertical fence of height, Γ, oriented along a straight line forming an angle, χ, from a north–south line, casts a shadow on a flat horizontal surface with a width, w, given by:

$$w = \Gamma \left| \sin (Z - \chi)/\tan A \right| \qquad (2\text{–}21)$$

where: w is the width of the shadow of a fence of height, Γ, directed at an angle, χ, from a north–south line

Z is the azimuth angle of the sun given by Equation 2–10 or 2–11.

Care must be taken to be sure that the two angles, Z and χ, are determined in the same sense from due south. A change in sign of the sine function in Equation 2–21 occurs as the shadow changes from one side of the fence to the other so the absolute value function is used to restore a positive value to w, regardless of which side of the fence is shaded.

The general treatment for finding the length and direction of a shadow cast by a tilted pole upon a tilted surface requires the use of a sequence of formulas. Consider a pole of length, Y, tilted from the zenith by an angle, ξ, in a direction, ψ, from due south. If its base is given as a plane surface tilted from the horizontal by an angle, Σ, in the direction, ϕ, from south, then the length of its shadow may be determined by using the following procedure:

Determine the angle, ζ, between the pole and the normal to the surface by:

$$\cos \zeta = \cos \Sigma \cos \xi + \sin \Sigma \sin \xi \cos (\psi - \phi) \qquad (2\text{–}22)$$

Then, the azimuthal angle, α, between the projection of the pole on the surface and the line of steepest slope is determined by:

$$\tan \left(\frac{\alpha}{2}\right) = \frac{\sin \xi \sin (\psi - \phi)}{\sin \zeta - \sin \Sigma \cos \xi + \cos \Sigma \sin \xi \cos (\psi - \phi)} \qquad (2\text{–}23)$$

Next, determine the incident angle, θ, of the sun to a line normal to the tilted plane surface by:

$$\cos \theta = \cos \Sigma \sin A + \sin \Sigma \cos A \cos (Z - \phi) \qquad (2\text{–}24)$$

where: A and Z are the elevation and azimuth, respectively, of the sun from Equations 2–9 and 2–10 or 2–9 and 2–11.

Then, determine the azimuthal sun-line angle, β, on the tilted surface measured from the direction of steepest tilt from:

$$\tan \left(\frac{\beta}{2}\right) = \frac{\cos A \sin (Z - \phi)}{\sin \theta - \sin \Sigma \sin A + \cos \Sigma \cos A \cos (Z - \phi)} \qquad (2\text{–}25)$$

Next, determine the orthogonal parameters:

$$M = \sin \zeta \sin \alpha - \cos \zeta \sin \beta \tan \theta \qquad \textbf{(2–26)}$$
$$N = \sin \zeta \cos \alpha - \cos \zeta \cos \beta \tan \theta \qquad \textbf{(2–27)}$$

The length of the shadow is:

$$l = Y \sqrt{M^2 + N^2} \qquad \textbf{(2–28)}$$

and its orientation on the plane will be at an angle, Ω, from the direction of steepest tilt determined by:

$$\Omega = \tan^{-1} \left(\frac{M}{N} \right) + 180° \qquad \textbf{(2–29)}$$

Although this sequence of equations seems complex, it is often surprising how quickly they reduce to simpler forms for the usual geometries which are normally encountered in solar energy applications.

For example, consider the case of a pole of length Y, normal to a tilted surface. In that case $\xi = \Sigma$, $\phi = \psi$, and from Equation 2–22, $\zeta = 0$. This causes Equations 2–26 and 2–27 to reduce to $M = -\sin \beta \tan \theta$ and $N = -\cos \beta \tan \theta$ which, in turn, results in the shadow length, $l = Y \tan \theta$, from Equation 2–28. Equation 2–29 provides the angle formed by the shadow: $\Omega = \tan^{-1} (\tan \beta) +$

TABLE 2–5   The Shadow of a Pole: Length and Direction

Vertical pole of height, Y, on horizonal plane:	$l = \dfrac{Y}{\tan A}$ $\Omega = Z + 180°$
Pole of length, Y, normal to plane tilted by Σ from horizontal in direction, ϕ, from south:	$l = Y \tan \theta$ $\Omega = \beta + 180°$ where $\sin \beta = \dfrac{\cos A \sin (Z - \phi)}{\sin \theta}$

Vertical pole of height, Y, on plane tilted by Σ from horizontal in direction, ϕ, from south:

$$l = Y \sqrt{\sin^2 \Sigma + \cos^2 \Sigma \tan^2 \theta - 2 \sin \Sigma \cos \Sigma \tan \theta \cos \beta}$$

$$\Omega = \tan^{-1} \left\{ \frac{\cos \Sigma \sin \beta \tan \theta}{\cos \Sigma \tan \theta \cos \beta - \sin \Sigma} \right\} + 180°$$

Pole, tilted by ξ in direction, ψ, from south on horizontal surface:

$$l = Y \sqrt{\sin^2 \xi + \frac{\cos^2 \xi}{\tan^2 A} - 2 \sin \xi \cos \xi \frac{\cos (Z - \psi)}{\tan A}}$$

$$\Omega = \tan^{-1} \left\{ \frac{\sin \xi \sin \psi \tan A - \cos \xi \sin Z}{\sin \xi \cos \psi \tan A - \cos \xi \cos Z} \right\} + 180°$$

Y = height (length) of pole
ξ = angle by which pole is tilted from vertical
ψ = direction from south by which pole is tilted
Σ = angle by which plane is tilted from horizontal
ϕ = angle from south in which plane is tilted
θ = incident angle of sun (Equation 2–24)
β = tilted surface azimuth of sun (Equation 2–25)
A = elevation angle of sun
Z = azimuth angle of sun

$180° = 180° + \beta$, as would be expected where β is determined through Equation 2–25.

The general treatment for finding the width of the shadow cast by a tilted fence on a tilted surface can now be considered with the formulas discussed here and repeated in Table 2–6.

Consider a fence of height, Γ, where "height" is the perpendicular extent of the fence from its base to its upper edge. Suppose it is oriented along a line forming an angle, χ, with the lines of steepest tilt on a tilted plane surface. The plane surface is tilted by an angle, Σ, from the horizontal in a direction, ϕ, from south. Furthermore, let the fence be tilted at an angle, ξ, from the normal to the plane.

The width of its shadow, measured from the base of the fence along a perpendicular upon the tilted surface is:

$$w = \Gamma|\cos \xi \sin (\beta - \chi)\tan \theta| \pm \Gamma \sin \xi \qquad (2\text{–}30)$$

where the "+" sign is used when the shadow is in the direction of tilt and a "−" sign is used when the shadow is directed opposite to the tilt. The incident angle, θ, and surface azimuth, β, are found from Equations 2–24 and 2–25.

Equation 2–30 reduces to other convenient forms when special cases are considered. If the fence is normal to the plane, $\xi = 0$, and:

$$w = \Gamma|\sin (\beta - \chi)\tan \theta| \qquad (2\text{–}31)$$

If the fence is oriented along the line of steepest tilt and normal to the plane, then $\xi = 0$ and $\chi = 0$ so that $w = \Gamma \sin \beta \tan \theta$ or, from Equation 2–25, it reduces to:

$$w = \Gamma \left|\frac{\cos A \sin (Z - \phi)}{\cos \theta}\right| \qquad (2\text{–}32)$$

Equation 2–32 can also be used to determine the width of the shadow cast by the side wall of a tilted solar collector onto its absorber surface.

If the fence is normal to the surface and oriented perpendicular to the line of steepest tilt on the plane surface, then $\xi = 0$, and $\chi = 90°$ so that $w = \Gamma \cos \beta \tan \theta$, and:

$$w = \Gamma \frac{\sqrt{|\sin^2 \theta - \cos^2 A \sin^2 (Z - \phi)|}}{\cos \theta} \qquad (2\text{–}33)$$

Equation 2–33 can be used to determine the width of the shadow cast by the end, top, or bottom wall of a tilted solar collector onto its absorber surface.

For a vertical fence on a horizontal surface, then $\xi = 0$, $\Sigma = 0$, $\beta = Z$, and $\theta = 90° - A$ so that $w = \Gamma|\sin (Z - \chi)/\tan A|$ as stated in Equation 2–21. This equation is particularly useful in establishing the minimum spacing between rows of solar collectors.

Typically, solar collectors may be aligned in rows that are in an east–west direction ($\chi = 90°$) and tilted toward the south by an angle, Σ. If regarded as a tilted fence on a horizontal surface, then $\xi = 90° - \Sigma$, $\theta = 90° - A$, $\beta = Z$, and $\chi = 90°$. For this important case the shadow width is:

$$w = \Gamma|\sin \Sigma \cos Z/\tan A| \pm \Gamma \cos \Sigma \qquad (2\text{–}34)$$

as measured from the base or foot of the row of collectors.

In the winter when the sun's elevation, A, is low and its azimuth, Z, moves within a low daily range, the shadow width is large. Note, however, that reducing the tilt angle, Σ, can reduce the shadow width. In the summer when

TABLE 2-6 The Width of the Shadow of a Fence

Vertical fence of height, Γ, on horizontal surface oriented along line at angle, χ, from south:	$w = \Gamma \left\| \dfrac{\sin(Z - \chi)}{\tan A} \right\|$
Fence normal to tilted plane. Plane tilted by Σ from horizontal in direction, ϕ, from south. Fence directed at angle χ from steepest slope on tilted plane:	$w = \Gamma \left\| \sin(\beta - \chi)\tan\theta \right\|$
Fence normal to tilted plane along line of steepest slope. Plane tilted by Σ in direction, ϕ:	$w = \Gamma \left\| \dfrac{\cos A \, \sin(Z - \phi)}{\cos\theta} \right\|$
Fence normal to tilted plane perpendicular to line of steepest slope. Plane tilted by Σ in direction, ϕ:	$w = \Gamma \dfrac{\sqrt{\left\|\sin^2\theta - \cos^2 A \, \sin^2(Z - \phi)\right\|}}{\cos\theta}$
Fence tilted from normal by angle ξ on tilted plane. Plane tilted by Σ in direction, ϕ. Fence oriented at angle χ from steepest slope:	$w = \Gamma \left\| \cos\xi \, \sin(\beta - \chi)\tan\theta \right\| \\ \pm \Gamma \sin\xi$

Γ = height (extent) of fence
χ = angle formed between base of fence and steepest slope
Σ = tilt angle of plane from horizontal
ϕ = angle from south by which plane is tilted
θ = incident angle (Equation 2–24)
β = tilted surface azimuth of sun (Equation 2–25)
A = elevation angle of sun
Z = azimuth angle of sun

the sun's elevation, A, is high and its daily azimuth range is high, the shadow width is reduced and rows of solar collectors could be placed closer together. In designing an array of solar collectors, therefore, the determination of the optimum tilt and spacing between rows requires consideration of the shadows cast by adjacent rows at different times of both the year and the day. This important consideration is illustrated in a sample calculation which follows.

COLLECTOR SPACING

As has been noted, the width of the shadow cast by a row of solar collectors is needed to determine the minimum spacing between adjacent rows. Because the "sawtooth" array is commonly used for multiple panel installations, calculations of the appropriate spacing between rows can be instructive. Consider a "sawtooth" array at a height of 6 feet above a level rooftop surface, and oriented in an east–west direction at an angle of 90° from south (such that the collectors will be facing due south). Assume a latitude of 40° north.

In this example, maximum shadow widths will occur on the winter solstice—on or about December 21. The azimuth, Z, and elevation, A, of the sun on that date, on the summer solstice, and at the equinoxes for various times of the solar day are shown in the following schedule. These values may be calculated from Equations 2–9 and 2–10 or are available from various references. Shadow width calculations uses $\Gamma = 6$ feet and $\chi = 90°$.

From the first formula in Table 2–6:

$$w = 6 \text{ feet} \times \frac{\sin(Z - 90°)}{\tan A} = 6\frac{\cos Z}{\tan A}$$

TABLE 2–7 Shadows at Solstices and Equinoxes

Solar Time		Sun's Elevation	Sun's Azimuth	Shadow Width (feet)
AM	PM	A	Z	w
At December 21 (winter solstice)				
8:00	4:00	5.5°	53.0°	37.5
9:00	3:00	14.0°	41.9°	17.9
10:00	2:00	20.7°	29.4°	13.8
11:00	1:00	25.0°	15.2°	12.4
12:00 noon		26.6°	0.0°	12.0
At June 21 (summer solstice)				
5:00	7:00	4.2°	117.3°	*
6:00	6:00	14.8°	108.4°	*
7:00	5:00	26.0°	99.7°	*
8:00	4:00	37.4°	90.7°	*
9:00	3:00	48.8°	80.2°	0.9
10:00	2:00	59.8°	65.8°	1.4
11:00	1:00	69.2°	41.9°	1.7
12:00 noon		73.4°	0.0°	1.8
At March 21 and September 21 (equinoxes)				
7:00	5:00	11.4°	80.2°	5.0
8:00	4:00	22.5°	69.6°	5.0
9:00	3:00	32.8°	57.2°	5.0
10:00	2:00	41.5°	41.9°	5.0
11:00	1:00	47.7°	22.6°	5.0
12:00 noon		50.0°	0.0°	5.0

* = negative values; the shadow is in front of the collector.

From this example, it is clear that at the winter solstice a minimal spacing of approximately 18 feet between adjacent rows would be necessary to accommodate winter solstice solar angles and shadow widths. Such a spacing would automatically accommodate shadow widths at other times of the year, since the shadow width decreases to 5 feet at the equinoxes (and is constant throughout the day), and is reduced to less than 2 feet on the summer solstice.

If spacing of 18 feet is used, however, a considerable loss of usable rooftop area would occur in the summer months with sunlight falling between adjacent rows. This is one of the unfortunate limitations in making optimal use of horizontal areas for solar energy collection. The most ideal "sawtooth" array geometry for one season cannot be effectively applied for all seasons.

A study of shadow widths also points out other problems inherent in the use of a "sawtooth" array. Ideally, the performance of such a collector array could be improved by tilting the collectors on a seasonal basis. But, if this done, the height of the collectors becomes higher in the winter, thus requiring even greater spacing between rows since shadow width will increase with increases in height. This has led some collector suppliers to use retro-reflectors on the back, north-facing side of the rows of a "sawtooth" array. These reflectors serve to increase the energy collection in the summer, since they collect some of the solar energy that would otherwise fall between adjacent rows.

CHAPTER 3
CLIMATIC STATISTICS AND
SOLAR INSOLATION

All the variations affecting the magnitude of solar intensity or, more correctly, solar insolation at the earth's surface which have been considered in Chapter 2 are regular in the sense that their recurring effects are predictable. At a given site, seasonal variations of sun and shadow will be reproduced from year to year, the effects of latitude are as unchangeable as latitude itself, and the universal waxing and waning of daily light has become the proverbial difference between night and day. But climate is capricious and its effects demand that each installation, even those in locations near enough or similar enough to tempt standardization, be treated individually. Even the smallest planning generalization is unwise because climatological peculiarities can yield surprising results. New York City, for example, is farther south in latitude than Rome, Italy, but because climate is a product of influences more complex than simple latitude, solar energy systems in New York must cope with greater temperature extremes and more cloudy days than those in Rome. Climate is, therefore, the dominant variable to be considered in the design of all solar energy systems. Of that factor there are three principal concerns: the percentage of sun versus cloud cover, the average ambient temperature, and both peak and average wind speeds.

SUN VERSUS CLOUD COVER

The percentage of sun versus cloud cover can be expressed in several ways. One might observe the sun over a long period and measure those intervals during which the sun is not obscured versus those when clouds interfere. Another approach, to measure the longest time period between successive appearances of the sun, if done over a time long enough to be statistically valid, could provide data useful in determining minimum solar energy storage requirements. The most complete and accurate approach is to record the average daily solar insolation (both direct and diffuse components) with a continuously operating solar radiometer. Figures 3–1 through 3–12 depict the results of this method by month for the United States as averaged over a period of twelve years.

The maps in Figures 3–1 through 3–12 were plotted by the National Oceanic and Atmospheric Administration (NOAA), using data from the old network of weather stations in the U.S. Unfortunately, the NOAA network had

(*text continued on page 47*)

33

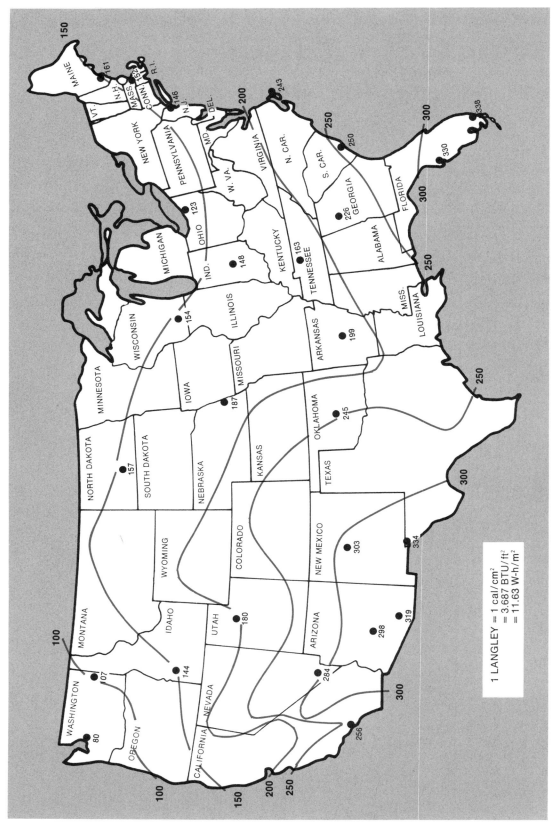

FIGURE 3-1 Average daily solar radiation, January (Langleys per day).

1 LANGLEY = 1 cal/cm²
= 3.687 BTU/ft²
= 11.63 W-h/m²

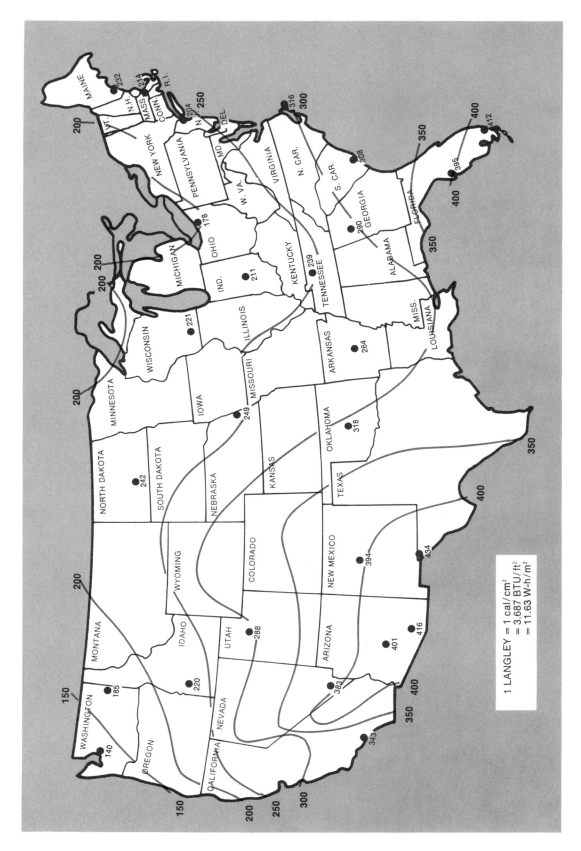

FIGURE 3-2 Average daily solar radiation, February (Langleys per day).

1 LANGLEY = 1 cal/cm²
= 3.687 BTU/ft²
= 11.63 W-h/m²

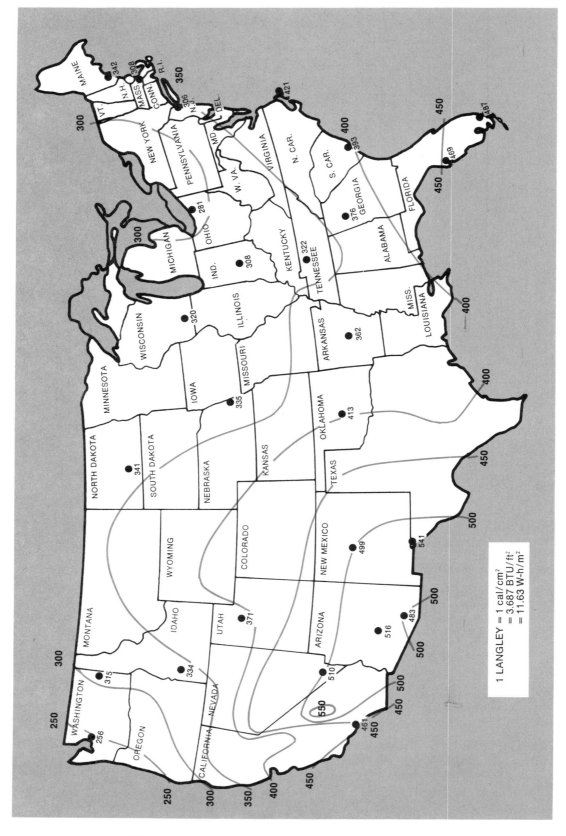

FIGURE 3-3 Average daily solar radiation, March (Langleys per day).

1 LANGLEY = 1 cal/cm²
= 3.687 BTU/ft²
= 11.63 W-h/m²

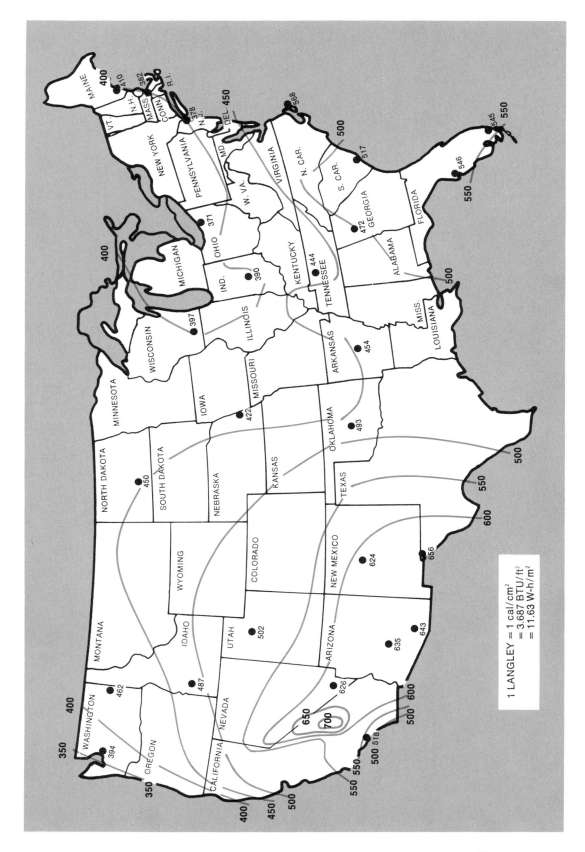

FIGURE 3-4 Average daily solar radiation, April (Langleys per day).

1 LANGLEY = 1 cal/cm²
= 3.687 BTU/ft²
= 11.63 W-h/m²

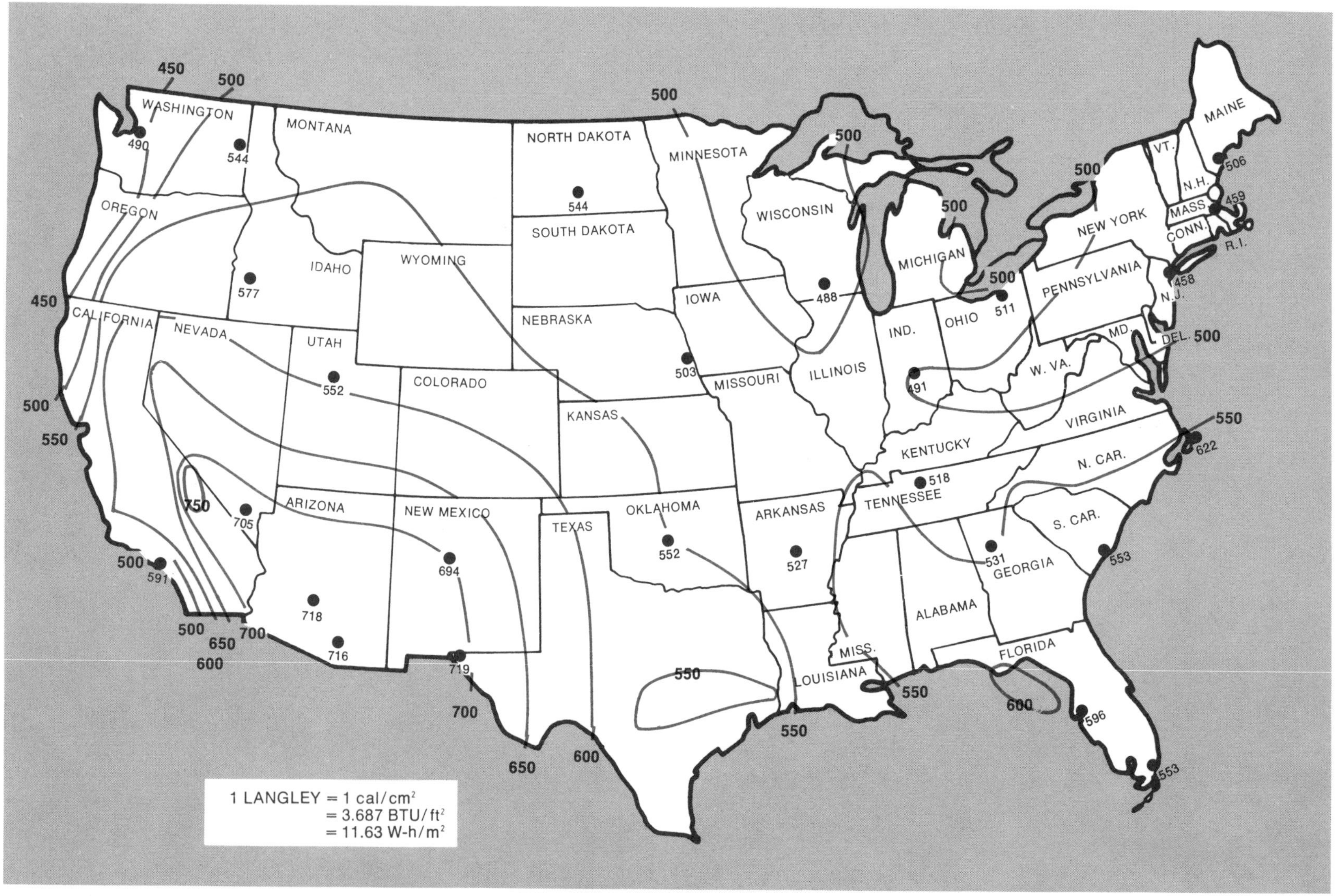

FIGURE 3-5 Average daily solar radiation, May (Langleys per day).

1 LANGLEY = 1 cal/cm²
= 3.687 BTU/ft²
= 11.63 W-h/m²

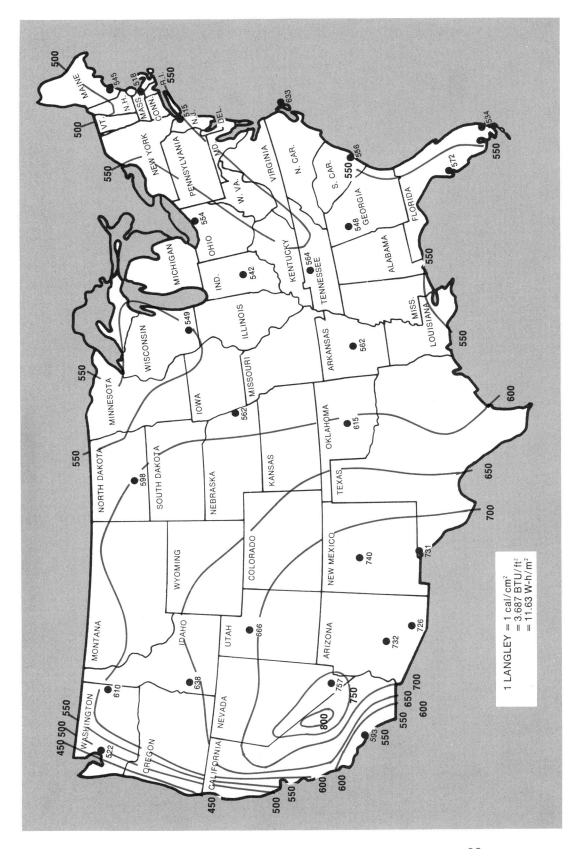

FIGURE 3-6 Average daily solar radiation, June (Langleys per day).

1 LANGLEY = 1 cal/cm²
= 3.687 BTU/ft²
= 11.63 W-h/m²

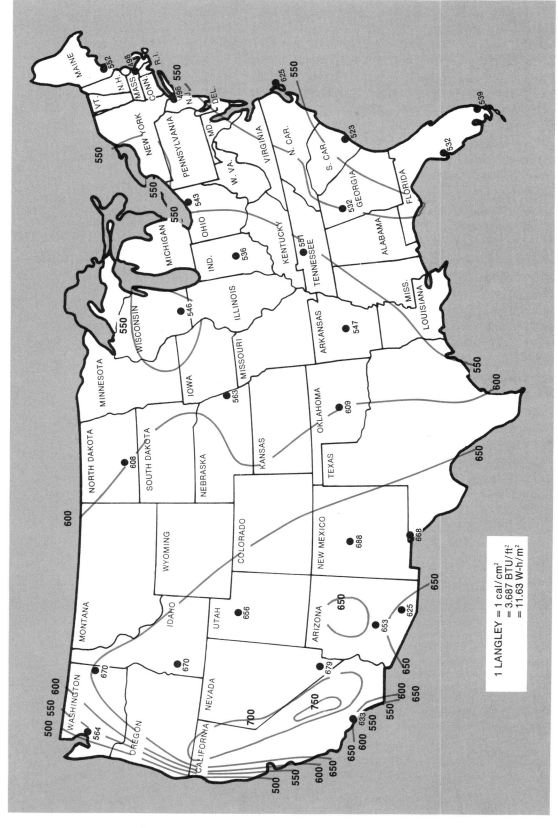

FIGURE 3-7 Average daily solar radiation, July (Langleys per day).

1 LANGLEY = 1 cal/cm²
= 3.687 BTU/ft²
= 11.63 W-h/m²

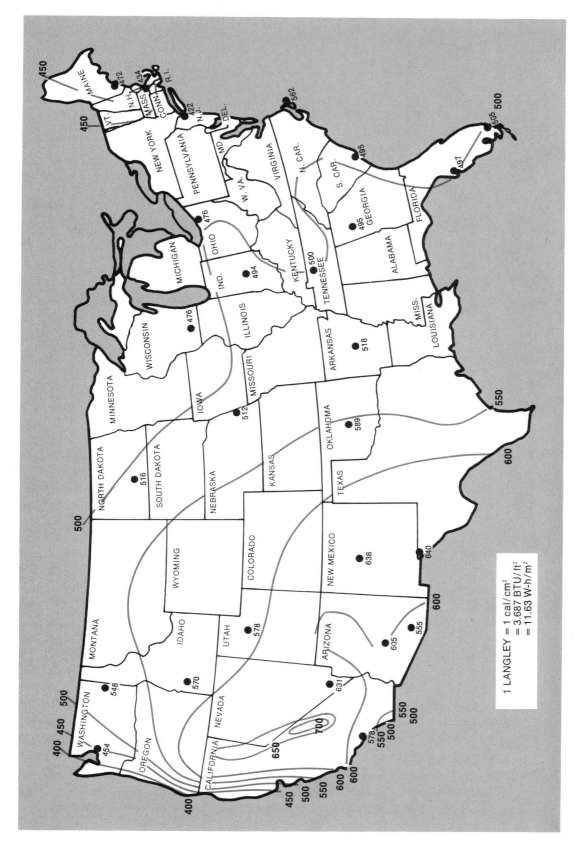

FIGURE 3-8 Average daily solar radiation, August (Langleys per day).

1 LANGLEY = 1 cal/cm²
         = 3.687 BTU/ft²
         = 11.63 W-h/m²

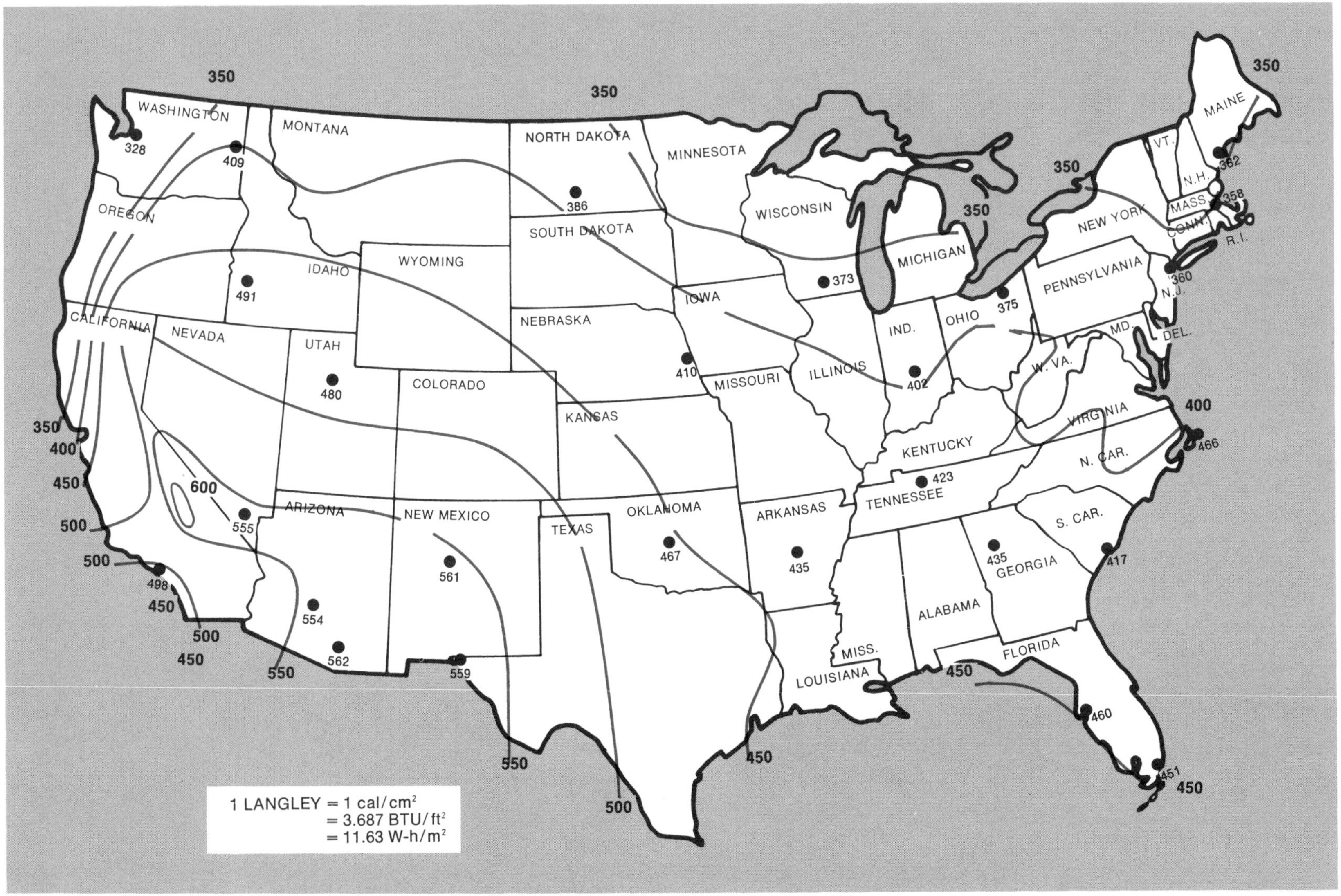

FIGURE 3-9 Average daily solar radiation, September (Langleys per day).

1 LANGLEY = 1 cal/cm²
= 3.687 BTU/ft²
= 11.63 W-h/m²

42

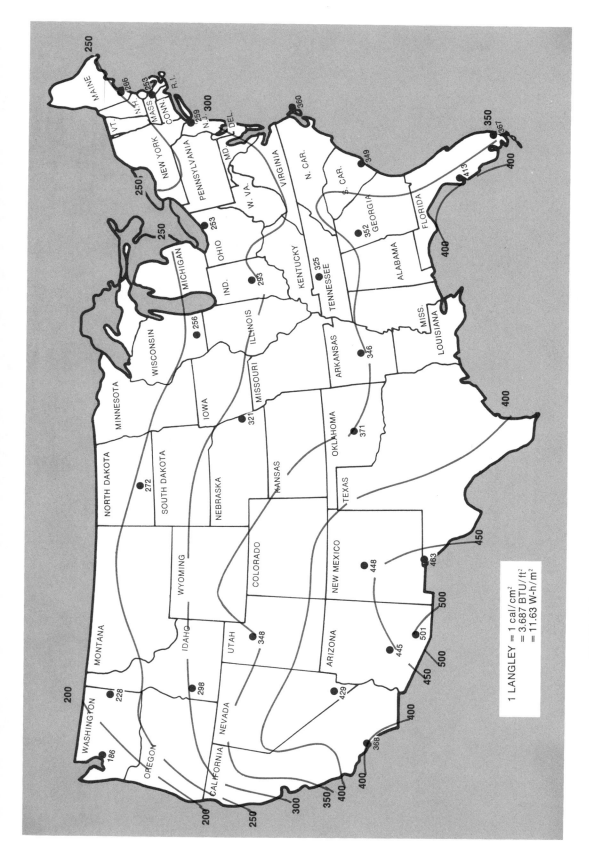

FIGURE 3–10 Average daily solar radiation, October (Langleys per day).

1 LANGLEY = 1 cal/cm²
= 3.687 BTU/ft²
= 11.63 W-h/m²

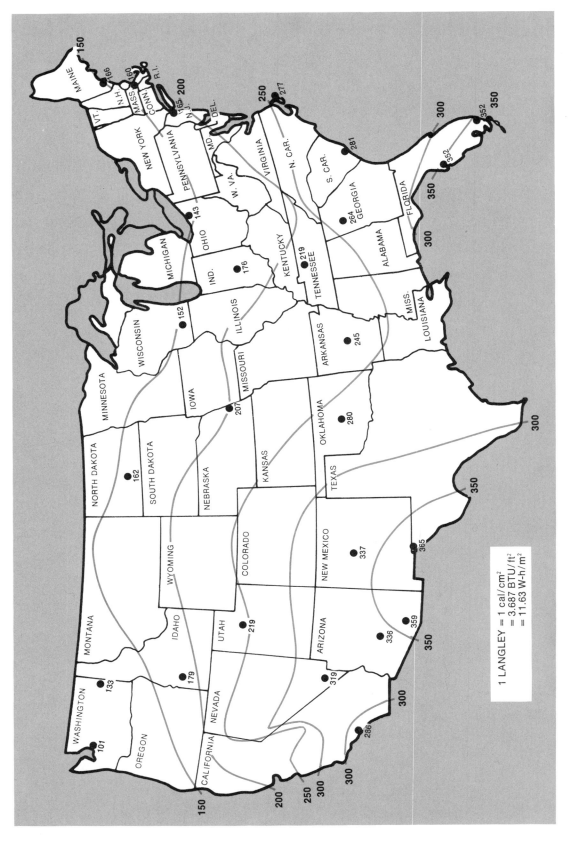

FIGURE 3-11 Average daily solar radiation, November (Langleys per day).

1 LANGLEY = 1 cal/cm² = 3.687 BTU/ft² = 11.63 W-h/m²

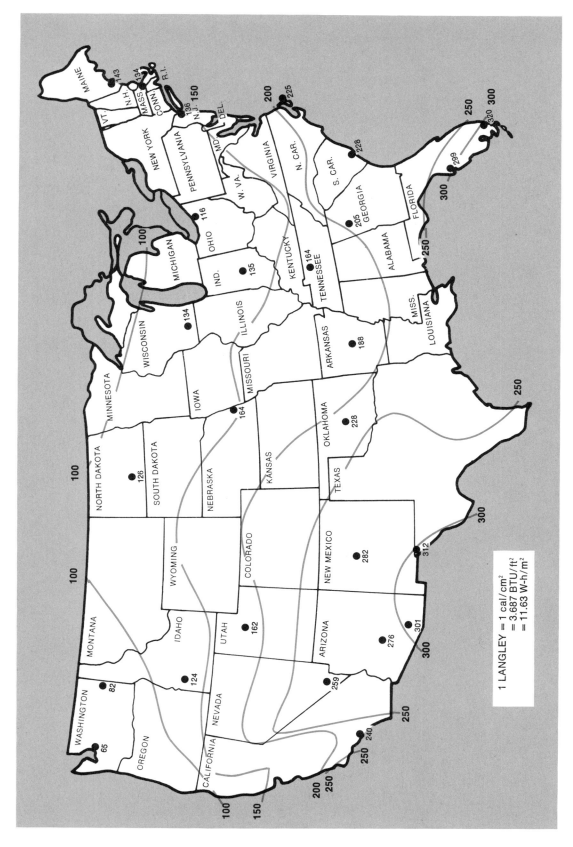

FIGURE 3-12 Average daily solar radiation, December (Langleys per day).

1 LANGLEY = 1 cal/cm²
= 3.687 BTU/ft²
= 11.63 W-h/m²

TABLE 3–1 Average Percentage of Possible Sunshine: Selected Cities

[Average data, except as noted. For period of record through 1975]

	State and Station	Length of record (Years)	Jan.	Feb.	Mar.	Apr.	May	June	July	Aug.	Sept.	Oct.	Nov.	Dec.	Annual
AL	Montgomery	25	47	53	58	64	66	65	63	65	63	66	57	50	59
AK	Juneau	30	33	32	37	38	38	34	30	30	25	19	23	20	31
AZ	Phoenix	80	78	80	83	89	93	94	85	85	89	88	83	77	86
AR	Little Rock	32	46	54	57	61	68	73	71	73	68	69	56	48	63
CA	Los Angeles	32	69	72	73	70	66	65	82	83	79	73	74	71	73
	Sacramento	27	45	61	70	80	86	92	97	96	94	84	64	46	79
	San Francisco	38	56	62	69	73	72	73	66	65	72	70	62	53	67
CO	Denver	26	72	71	69	66	64	70	70	72	75	73	65	68	70
CN	Hartford	21	58	57	56	57	58	58	61	63	59	58	46	48	57
DE	Wilmington*	25	50	54	57	57	59	64	63	61	60	60	54	51	53
DC	Washington	27	48	51	55	56	58	64	62	62	62	60	53	47	57
FL	Jacksonville	25	57	61	66	71	69	61	59	58	53	56	61	56	61
	Key West	17	72	76	81	84	80	71	75	76	69	68	72	74	75
GA	Atlanta	41	47	52	57	65	69	67	61	65	63	67	60	50	61
HI	Honolulu	23	63	65	69	67	70	71	74	75	75	67	60	59	68
ID	Boise	35	41	52	63	68	71	75	89	85	82	67	45	39	67
IL	Chicago	33	44	47	51	53	61	67	70	68	63	62	41	38	57
	Peoria	32	45	50	52	55	59	66	68	67	64	63	44	39	57
IN	Indianapolis	32	41	51	51	55	61	68	70	71	66	64	42	39	58
IO	Des Moines	25	51	54	54	55	60	67	71	70	64	64	49	45	59
KS	Wichita	22	59	59	60	62	64	69	74	73	65	66	59	56	65
KY	Louisville	28	41	47	50	55	62	67	66	68	65	63	47	39	57
LA	Shreveport	23	49	54	56	55	64	71	74	72	68	71	62	53	64
ME	Portland	35	55	59	56	56	56	60	64	65	61	58	47	53	58
MD	Baltimore	25	51	55	55	55	57	62	65	62	60	59	51	48	57
MA	Boston	40	54	56	57	56	58	63	66	67	63	61	51	52	59
MI	Detroit	32	32	43	49	52	59	65	70	65	61	56	35	32	54
	Sault Ste. Marie	34	34	46	55	55	56	57	63	58	46	41	23	28	48
MN	Duluth	25	49	54	56	54	55	58	67	61	52	48	34	39	54
	Minneapolis-St. Paul	37	51	57	54	55	58	63	70	67	61	57	39	40	58
MS	Jackson	11	48	55	61	60	63	67	61	62	58	65	54	45	59
MO	Kansas City	3	64	54	61	65	67	72	84	69	51	62	46	54	64
	St. Louis	16	52	51	54	56	62	69	71	66	63	62	49	41	58
MT	Great Falls	33	49	57	67	62	64	65	81	78	68	61	46	46	64
NE	Omaha	40	55	55	55	59	62	68	76	72	67	67	52	48	62
NV	Reno	33	66	68	74	80	81	85	92	93	92	83	70	63	80
NH	Concord	34	52	54	52	53	54	57	62	60	54	54	42	47	54
NJ	Atlantic City	15	49	48	51	53	54	58	60	62	59	57	50	42	54
NM	Albuquerque	36	73	73	74	77	80	83	76	76	80	79	78	72	77
NY	Albany	37	46	51	52	53	55	59	64	61	56	53	36	38	53
	Buffalo	32	34	40	46	52	58	66	69	66	60	53	29	27	53
	New York†	99	50	55	56	59	61	64	65	64	63	61	52	49	59
NC	Charlotte	25	55	59	63	70	69	71	68	70	68	69	63	58	66
	Raleigh	21	55	58	63	64	60	61	61	61	60	63	63	56	60
ND	Bismarck	36	54	56	60	58	63	64	76	73	65	59	44	47	62
OH	Cincinnati	60	41	45	51	55	61	67	68	67	66	59	44	38	57
	Cleveland	34	32	37	44	53	59	65	68	64	60	55	31	26	52
	Columbus	24	37	41	44	52	58	62	64	63	62	58	38	30	53
OK	Oklahoma City	23	59	61	63	63	65	73	75	77	69	68	60	59	67
OR	Portland	26	24	35	42	48	54	51	69	64	60	40	27	20	47
PA	Philadelphia	33	50	53	56	56	57	63	63	63	60	60	53	49	58
	Pittsburgh	23	36	38	45	48	53	60	62	60	60	56	40	30	50
RI	Providence	22	57	56	55	55	57	57	59	59	58	60	49	51	56
SC	Columbia	22	56	59	64	67	66	65	64	65	65	66	64	60	63
SD	Rapid City	33	54	59	61	59	57	60	71	73	67	65	56	54	62
TN	Memphis	25	48	54	57	63	69	73	72	75	69	71	58	49	64
	Nashville	33	40	47	52	59	62	67	64	66	63	64	50	40	57
TX	Amarillo	34	69	68	71	73	73	77	77	78	74	75	73	67	73
	El Paso	33	78	82	85	87	89	89	79	80	82	84	83	78	83
	Houston	6	41	54	48	51	57	63	68	61	57	61	58	69	56
UT	Salt Lake City	38	47	55	64	66	73	78	84	83	84	73	54	44	70
VT	Burlington	32	42	48	52	50	56	60	65	62	55	50	30	33	51
VA	Norfolk	19	57	58	63	66	67	68	65	65	64	60	60	57	63
	Richmond	25	51	54	59	62	64	67	65	64	63	59	56	51	60
WA	Seattle-Tacoma	10	21	42	49	51	58	54	67	65	61	42	27	17	49
	Spokane	27	26	41	53	60	63	65	81	78	71	51	28	20	57
WV	Parkersburg	78	32	36	43	49	56	59	62	60	59	54	37	29	48
WI	Milwaukee	35	44	47	51	54	59	63	70	67	60	57	41	38	56
WY	Cheyenne	40	61	65	64	61	59	65	68	68	69	68	60	59	64
PR	San Juan	20	65	69	74	69	61	57	64	65	59	59	57	56	63

Source: U.S. National Oceanic and Atmospheric Administration, *Comparative Climatic Data.*
* Data not available; figures are for a nearby station.
† City office data.

several deficiencies both in coverage and accuracy of the data. Beginning in 1973, efforts were made to correct the situation with 38 new stations now operational. The NOAA will be releasing new maps based on this more accurate data as they are prepared; as of March 1979, data and maps for January and February of 1977 are available. In addition, weather tapes for selected sites using rehabilitated data are now being supplied in Fortran magnetic tape.

The maps in Figures 3–1 through 3–12, therefore, are useful for illustration and sample calculations; analysis of a specific application site, and calculations for sizing a solar energy system, should utilize the new, accurate data. The work by Liu and Jordan in correlating diffuse radiation with total radiation is helpful in predicting long-term performance. Whillier, in his doctoral thesis, and Klein have refined this method further.

A numerical representation of the percentage of sun versus cloud cover is given for selected cities in Table 3–1; this specific information will be used in determining load requirements through the solar engineering worksheets. For the moment, it is important to recognize that cloud effects are not limited to blockage of the sun with a consequent diminution in the magnitude of solar insolation. The presence of clouds not directly obscuring the sun can have effects ranging from zero to a significant increase in the level of solar insolation resulting from the increase in diffuse radiation by cloud scattering. Accordingly, it is necessary to refine the calculation of total solar insolation by considering the effects of all its components.

TOTAL SOLAR INSOLATION

The total solar insolation incident upon a surface consists of three components. The first is direct-beam radiation, following a direct path from the sun to the object and forming an angle θ with the normal to the surface. The second component is diffuse radiation, sunlight which has been scattered by molecules and particles in the atmosphere (such as the moisture in clouds) and which arrives from all directions of the sky. The last component is the radiation of sunlight reflected by the ground and objects on the ground.

Total solar insolation incident to a flat surface is given by:

$$I_T = I_b + I_d + I_r \qquad \text{(3–1)}$$

where: I_T is the total solar insolation incident to the surface in BTU/hr-ft²

I_b is the direct beam component of solar insolation given by Equations 2–13 and 2–19 in BTU/hr-ft²

I_d is the diffuse component of solar insolation (BTU/hr-ft²)

I_r is the reflected component of solar insolation (BTU/hr-ft²).

On clear days (i.e., cloudless and sunny), the diffuse component, I_d, and the reflected component, I_r, of solar insolation are approximated by:

$$I_d = C_d I_n F_s \qquad \text{(3–2)}$$
$$I_r = C_r I_n F_g \qquad \text{(3–3)}$$

where: $\left. \begin{aligned} F_s &= (1 + \cos \Sigma)/2 \\ F_g &= (1 - \cos \Sigma)/2 \end{aligned} \right| \qquad \text{(3–4)}$

TABLE 3–2   Coefficients for Diffuse and Reflected Solar Insolation
on a Clear Day

Month	C_d	Surface	C_r
Jan.	.058	Snow	0.7–0.87
Feb.	.060		
Mar.	.071	Bituminous and	
Apr.	.097	gravel roofs	0.12–0.15
May	.121		
June	.134	Bituminous	
July	.136	paving	0.10
Aug.	.122		
Sept.	.092	Concrete	0.21–0.33
Oct.	.073		
Nov.	.063	Grass	0.20–0.30
Dec.	.057		

are the weighted view factors for the sky and ground, respectively. C_d and C_r are given in Table 3–2 and the angle, Σ, is the angle by which the flat surface is tilted from the horizontal, as shown in Figure 2–6. I_n is the direct normal solar insolation given by Equation 2–14.

The coefficient for diffuse solar insolation, C_d, may also be represented by the formula:

$$C_d = 0.094 - 0.031 \cos y - 0.013 \sin \frac{3y}{2} \qquad \text{(3–5)}$$

The maps in Figures 3–1 through 3–12 are used to provide the average daily solar insolation in Langleys at any selected site within the continental United States. The data is averaged over each day of the month represented by the figure.

The value of solar insolation in Langleys is converted to BTUs per square foot per day by multiplying by 3.69. The result is the average total daily solar insolation available on a horizontal surface, I_H, in BTU/day-ft².

Diffuse radiation on clear days can be approximated by Equation 3–2; however, the average solar insolation data on the maps of Figures 3–1 through 3–12 are based on a composite of clear and cloudy days, such as they naturally occur. The nature and intensity of diffuse radiation on cloudy and partially cloudy days is quite different from that of clear days.

A fully equipped U.S. weather station will supply data on the cloud cover (CC) at the site of the station on an hourly basis for each day of the year. The cloud cover, CC, is usually estimated visually by experienced cloud observers and expressed as a number from 0 to 10. A clear sky with no clouds is designated as $CC = 0$ and a completely overcast sky is denoted as $CC = 10$.

A useful expression for hourly diffuse radiation intensity is provided by Kreider and Kreith:

$$I_{Hd} = 0.78 + 1.07A + 6.17CC \qquad \text{(3–6)}$$

where: I_{Hd} is the diffuse component of solar insolation on a horizontal surface in BTU/hr-ft²

A is the sun's angle of elevation in degrees, given by Equation 2–9

CC is the cloud cover (from 0 to 10).

The total daily value of diffuse radiation must be obtained by summing the hourly values of I_{Hd} obtained in Equation 3–6 through the hours of the day. Both A and CC change with each hour of the day.

If cloud cover data are available, the average daily value of diffuse solar insolation on a horizontal surface I_{Hd} in BTU/day-ft² can be obtained by a summation of hourly values in Equation 3–6.

The value of I_H, the total insolation on a horizontal surface is:

$$I_H = I_{Hb} + I_{Hd} \text{ (BTU/hr-ft}^2) \tag{3–7}$$

where: I_{Hd} is given on an hourly basis by Equation 3–6

I_{Hb} may be determined from the calculated values of the direct normal solar insolation and the air mass from Equation 2–19, but should be set = 0 when the sun is not visible.

If cloud cover data are not available, the measured daily values of solar insolation on a horizontal surface, I_H (BTU/day-ft²), may be obtained by multiplying the value in Langleys obtained from the maps of Figures 3–1 through 3–12 by 3.69, and the average daily values of I_{Hb} can be approximated by adding the hourly contributions obtained from Equation 2–19 and multiplying the result by the "percent sun" (%/100) value taken from Table 3–1. The average daily diffuse radiation on a horizontal surface, I_{Hd}, may then be represented as $I_H - I_{Hb}$ where the units are in BTU/day-ft².

In summary, the total hourly solar insolation on a tilted surface consists of direct, diffuse, and reflected components, as indicated in Equation 3–1. The values of the diffuse and reflected radiation for clear days are given by Equations 3–2 and 3–3 in terms of the weighted view factors that depend on the tilt angle, Equation 3–4, the coefficients C_d and C_r, and the direct normal solar insolation, I_n, which is provided in Equation 2–14.

On cloudy days the value of diffuse radiation is based on the cloud cover, CC, by Equation 3–6, and the value of beam radiation should be set = 0 whenever the sun is not visible.

The average daily solar insolation is shown on the maps in Figures 3–1 through 3–12; the percent sun figures in Table 3–1 can be used in conjunction with these maps to determine the average daily direct and diffuse radiation on a horizontal surface.

In many solar applications, one begins by assuming clear day performance and then multiplies the clear day energy delivered by the percent sun figure to obtain the average daily energy delivered. This is preferable to first assuming performance based on average values of solar insolation because the performance of a solar system is not linearly dependent upon solar insolation, with the effect that there is a certain value of solar insolation below which no net energy can be delivered by most systems. The percent sun figure best represents the proportion of days within any month that a solar system can be expected to be exposed directly to the sun—the most critical figure, since solar energy collection systems deliver zero net energy when the sun is not visible, and diffuse radiation by itself is of little value.

One advantage of a flat plate collector, however, is that it can benefit from the incremental addition of diffuse radiation to direct beam radiation when the sun is visible. This means that if hourly cloud cover information for use in Equation 3–6 is not available, there will be a tendency for the percent sun approach to understate, to some extent, the potential performance of a solar energy system using flat plate collectors. Systems that are based on concentrators which track the sun utilize only the direct normal component of solar insolation, I_n; they cannot collect energy from diffuse radiation.

AMBIENT TEMPERATURE

Ambient temperature is the environmental temperature in which a solar energy system operates; it is probably the single most important climatic variable to be considered in the design of that solar energy system. The effect of ambient temperatures through the duration of a season is most conveniently expressed in terms of "degree days" of heating or cooling requirements. The degree-day heating requirement is determined by totaling the daily average difference between 65°F and the average ambient temperature; degree-day cooling requirements are determined by totaling the daily average difference between ambient temperature and 75°F. Seasonal degree-day heating and cooling requirements by regional climatic classification and the average daily solar insolation for heating and cooling seasons are shown in Figures 3–13 and 3–14.

Degree-day information by month for representative cities, data with the precision necessary for use in the worksheets, is given in Table 3–3. Cooling degree-day data can be used in like calculations, but the information is not as accurate in determining air-conditioning energy loads since much of the energy is required to reduce humidity rather than to cool the air. For this reason, the average relative humidity in the same representative cities for each month of the year is provided in Table 3–4.

In the main, a solar system is engineered around the average monthly temperature at the site (shown for selected cities in Table 3–5). Excess thermal energy may be stored during periods when ambient is above this average, utilized for service when ambient falls below the average. It may be preferable, to ensure more complete fulfillment of energy requirements, to design the system around the normal daily *minimum* temperatures as shown in Table 3–6. Systems using the monthly minimum temperatures as the critical parameter will have a greater capacity during those low temperature periods but, as a consequence, they must also be capable, through storage or rejection mechanisms, of coping with the excess energy achieved at the normal daily maximum temperatures shown in Table 3–7. Finally, for purposes of comparison and to provide extreme points for the design of a solar energy system, Tables 3–8 and 3–9 contain the record high and low temperatures for selected cities.

WIND EFFECTS

Wind must be considered during the design of solar energy systems since it directly affects the collection equipment and the building which it serves, and

(*text continued on page 62*)

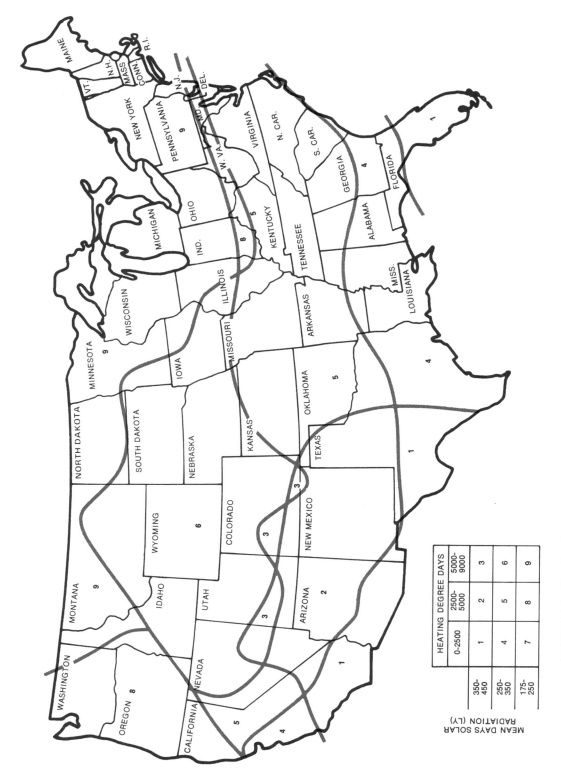

	HEATING DEGREE DAYS		
	0-2500	2500-5000	5000-9000
350-450	1	2	3
250-350	4	5	6
175-250	7	8	9

MEAN DAYS SOLAR
RADIATION (LY)

FIGURE 3–13 Regional climatic classification for the heating season (November–April).

51

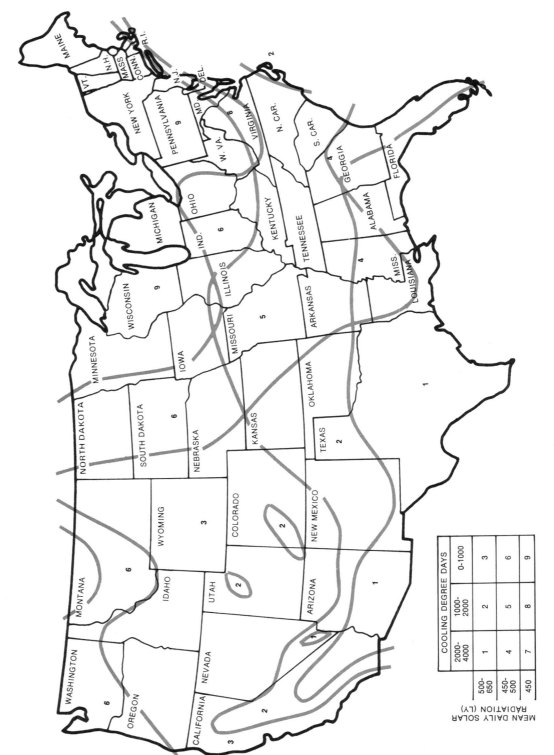

FIGURE 3-14 Regional climatic classification for the cooling season (May–October).

	COOLING DEGREE DAYS		
	2000-4000	1000-2000	0-1000
500-650	1	2	3
450-500	4	5	6
450	7	8	9

MEAN DAILY SOLAR
RADIATION (LY)

TABLE 3–3 Normal Monthly and Seasonal Heating Degree Days, 65° Base: Selected Cities

(Airport data, except as noted. Based on standard 30-year period, 1941 through 1970.)

State and Station		Jan.	Feb.	Mar.	Apr.	May	June	July	Aug.	Sept.	Oct.	Nov.	Dec.	Annual
AL	Mobile	451	337	221	40	–	–	–	–	–	39	211	385	1,684
AK	Juneau	1,287	1,036	1,026	783	564	354	288	332	474	719	975	1,169	9,007
AZ	Phoenix	428	292	185	60	–	–	–	–	–	17	182	388	1,552
AR	Little Rock	791	619	470	139	21	–	–	–	5	143	441	725	3,354
CA	Los Angeles	331	270	267	195	114	71	19	15	23	77	158	267	1,819
	Sacramento	617	426	372	227	120	20	–	–	5	101	360	595	2,843
	San Francisco	518	386	372	291	210	120	93	84	66	137	291	474	3,042
CO	Denver	1,088	902	868	525	253	80	–	–	120	408	768	1,004	6,016
CN	Hartford	1,246	1,070	911	519	226	24	–	12	106	384	711	1,141	6,350
DE	Wilmington	1,023	879	725	381	128	–	–	–	32	254	579	939	4,940
DC	Washington	911	776	617	265	72	–	–	–	14	190	510	856	4,211
FL	Jacksonville	348	282	176	24	–	–	–	–	–	19	161	317	1,327
	Miami	53	67	17	–	–	–	–	–	–	–	13	56	206
GA	Atlanta	701	560	443	144	27	–	–	–	8	137	408	667	3,095
HI	Honolulu	–	–	–	–	–	–	–	–	–	–	–	–	–
ID	Boise	1,116	826	741	480	252	97	–	12	127	406	756	1,020	5,833
IL	Chicago	1,262	1,053	874	453	208	26	–	8	57	316	738	1,132	6,127
	Peoria	1,277	1,044	859	416	180	17	–	8	70	327	753	1,147	6,098
IN	Indianapolis	1,150	960	784	387	159	11	–	5	63	302	699	1,057	5,577
IO	Des Moines	1,414	1,142	964	465	186	26	–	13	94	350	816	1,240	6,710
KS	Wichita	1,045	804	671	275	90	7	–	–	32	211	606	946	4,687
KY	Louisville	983	818	661	286	105	5	–	–	35	241	600	911	4,645
LA	New Orleans	403	299	188	29	–	–	–	–	–	40	179	327	1,465
ME	Portland	1,349	1,179	1,029	669	381	106	27	55	200	493	792	1,218	7,498
MD	Baltimore	980	846	688	340	110	–	–	–	27	250	567	921	4,729
MA	Boston	1,110	969	834	492	218	27	–	8	76	301	594	992	5,621
MI	Detroit	1,225	1,067	918	507	238	26	–	11	80	342	717	1,097	6,228
	Sault Ste. Marie	1,575	1,394	1,271	804	496	200	96	125	291	583	966	1,392	9,193
MN	Duluth	1,751	1,481	1,287	792	484	194	67	104	318	611	1,098	1,569	9,756
	Minneapolis-St. Paul	1,637	1,358	1,138	597	271	65	11	21	173	472	978	1,438	8,159
MS	Jackson	569	442	313	74	6	–	–	–	–	91	301	504	2,300
MO	Kansas City	1,153	893	745	314	111	12	–	–	42	235	642	1,014	5,161
	St. Louis	1,045	837	682	272	103	10	–	–	35	224	600	942	4,750
MT	Great Falls	1,380	1,075	1,070	648	367	162	18	42	260	524	912	1,194	7,652
NE	Omaha	1,314	1,036	865	391	148	20	–	6	71	301	750	1,147	6,049
NV	Reno	1,026	781	766	546	328	145	17	50	168	456	747	992	6,022
NH	Concord	1,376	1,187	1,014	624	315	58	16	45	182	487	810	1,246	7,360
NJ	Atlantic City	1,001	871	741	399	131	9	–	–	35	262	570	927	4,946
NM	Albuquerque	924	700	595	282	58	–	–	–	7	218	615	893	4,292
NY	Albany	1,349	1,162	980	543	253	39	9	22	135	422	762	1,212	6,888
	Buffalo	1,280	1,137	1,020	603	321	58	12	33	138	419	756	1,150	6,927
	New York*	1,017	885	741	387	137	–	–	–	29	209	528	915	4,848
NC	Charlotte	710	588	461	145	34	–	–	–	10	152	420	698	3,218
	Raleigh	760	638	502	180	48	–	–	–	12	186	450	738	3,514
ND	Bismarck	1,761	1,442	1,237	660	339	122	18	35	252	564	1,083	1,531	9,044
OH	Cincinnati	1,020	857	692	307	118	7	–	–	37	245	612	949	4,844
	Cleveland	1,181	1,039	896	501	244	40	9	17	95	354	702	1,076	6,154
	Columbus	1,135	972	800	418	176	13	–	8	76	342	699	1,063	5,702
OK	Oklahoma City	874	664	532	180	36	–	–	–	12	148	474	775	3,695
OR	Portland	834	622	598	432	264	128	48	56	119	347	591	753	4,792
PA	Philadelphia	1,014	871	716	367	122	–	–	–	38	249	564	924	4,865
	Pittsburgh	1,144	1,000	834	444	208	26	7	16	98	372	711	1,070	5,930
RI	Providence	1,135	997	871	531	259	36	–	10	93	350	651	1,039	5,972
SC	Columbia	608	493	360	83	12	–	–	–	–	112	341	589	2,598
SD	Sioux Falls	1,575	1,277	1,085	567	259	65	10	18	165	465	957	1,395	7,838
TN	Memphis	760	594	457	131	22	–	–	–	7	142	423	691	3,227
	Nashville	828	672	524	176	45	–	–	–	10	180	498	763	3,696
TX	Dallas-Fort Worth	626	456	335	88	–	–	–	–	–	60	287	530	2,382
	El Paso	663	465	328	89	–	–	–	–	–	92	402	639	2,678
	Houston	416	294	189	23	–	–	–	–	–	24	155	333	1,434
UT	Salt Lake City	1,147	885	787	474	237	88	–	5	105	402	777	1,076	5,983
VT	Burlington	1,494	1,299	1,113	660	331	63	20	49	191	502	840	1,314	7,876
VA	Norfolk	760	661	532	226	53	–	–	–	9	141	402	704	3,488
	Richmond	853	717	569	226	64	–	–	–	21	203	480	806	3,939
WA	Seattle-Tacoma	831	636	648	489	313	167	80	82	170	397	612	760	5,185
	Spokane	1,228	918	853	567	327	144	21	47	196	533	885	1,116	6,835
WV	Charleston	946	798	642	287	113	10	–	–	46	267	588	893	4,590
WI	Milwaukee	1,414	1,190	1,042	609	348	90	15	36	140	440	855	1,265	7,444
WY	Cheyenne	1,190	1,008	1,035	669	394	156	22	31	225	530	885	1,110	7,255
PR	San Juan	–	–	–	–	–	–	–	–	–	–	–	–	–

Source: U.S. National Oceanic and Atmospheric Administration, *Comparative Climatic Data.*
Note:–Represents zero.
* City office data.

TABLE 3–4 Average Relative Humidity, Percent: Selected Cities

(Airport data, except as noted. Eastern standard time. For period of record through 1975. Hours selected to give, for most of country, approximation of average highest and average lowest humidity values. Relative humidity observations were made on the half-hour prior to 1957.)

	State and Station	Length of Record (years)	Jan. 7:00 a.m.	Jan. 1:00 p.m.	Feb. 7:00 a.m.	Feb. 1:00 p.m.	Mar. 7:00 a.m.	Mar. 1:00 p.m.	Apr. 7:00 a.m.	Apr. 1:00 p.m.	May 7:00 a.m.	May 1:00 p.m.	June 7:00 a.m.	June 1:00 p.m.
AL	Mobile	13	82	64	81	56	84	56	87	55	86	53	86	54
AK	Juneau	32	79	76	81	75	79	69	75	64	74	63	75	64
AZ	Phoenix	15	44	30	37	25	33	23	23	15	17	12	18	12
AR	Little Rock	15	82	62	79	57	78	56	82	56	87	57	86	54
CA	Los Angeles	16	54	59	58	62	61	66	60	63	65	66	70	68
	Sacramento	15	86	71	79	61	68	52	58	43	52	37	48	32
	San Francisco	16	79	67	75	65	70	63	65	60	65	61	65	60
CO	Denver	15	45	48	44	43	42	41	38	34	38	37	38	37
CN	Hartford	16	72	57	73	56	72	53	69	44	74	47	79	53
DE	Wilmington	28	75	61	75	58	74	53	73	50	76	53	79	54
DC	Washington	15	69	54	68	53	68	49	68	48	72	51	76	53
FL	Jacksonville	39	88	57	85	53	85	49	85	48	84	49	86	56
	Miami	11	84	61	82	56	82	56	80	54	83	60	87	67
GA	Atlanta	15	80	61	76	55	79	52	80	52	83	55	86	59
HI	Honolulu	6	80	64	77	60	75	61	70	59	67	55	66	54
ID	Boise	36	73	71	68	61	55	45	47	36	45	34	42	31
IL	Chicago	12	73	65	72	62	74	59	74	56	73	53	75	54
	Peoria	16	78	69	79	66	82	64	78	57	81	57	82	57
IN	Indianapolis	16	80	69	79	66	79	63	77	55	82	57	82	58
IO	Des Moines	14	78	71	80	69	80	64	79	58	79	57	82	58
KS	Wichita	22	80	64	79	60	76	54	77	52	83	55	83	53
KY	Louisville	15	76	64	76	62	76	59	75	53	83	56	84	57
LA	New Orleans	27	86	67	85	63	85	61	88	61	89	60	89	62
ME	Portland	35	77	62	77	60	75	59	74	55	75	58	79	61
MD	Baltimore	22	72	58	72	56	71	51	72	49	77	52	80	53
MA	Boston	11	67	57	68	58	68	58	67	54	71	58	74	60
MI	Detroit	42	78	69	78	65	77	60	73	53	71	51	74	53
	Sault Ste. Marie	34	82	76	82	72	83	68	80	61	79	56	85	62
MN	Duluth	14	76	69	75	64	78	63	77	58	76	55	82	61
	Minneapolis-St. Paul	16	74	67	74	65	77	64	76	55	77	54	80	55
MS	Jackson	12	87	66	86	59	87	57	90	57	91	56	90	55
MO	Kansas City	3	75	65	75	65	78	62	74	54	80	56	81	54
	St. Louis	15	82	64	80	61	81	57	78	54	82	56	84	57
MT	Great Falls	14	62	61	59	55	53	47	50	42	46	40	45	40
NE	Omaha	12	76	66	77	62	76	57	76	53	79	54	81	56
NV	Reno	12	66	50	57	38	46	32	38	28	33	25	34	24
NH	Concord	10	74	59	75	59	78	57	77	47	80	49	88	57
NJ	Atlantic City	11	75	57	77	57	77	55	76	51	79	57	85	60
NM	Albuquerque	15	49	37	43	32	32	23	25	17	22	16	23	17
NY	Albany	10	80	63	77	58	73	54	69	47	74	52	78	56
	Buffalo	15	78	72	79	70	80	68	76	58	76	56	78	57
	New York*	55	68	60	68	58	67	55	68	51	71	53	74	55
NC	Charlotte	15	79	57	76	52	79	50	79	48	84	53	86	57
	Raleigh	11	78	56	74	49	79	49	80	45	86	55	88	57
ND	Bismarck	16	73	67	75	68	78	62	79	54	79	50	84	54
OH	Cleveland	15	76	69	77	68	77	65	75	57	77	58	79	59
	Columbus	16	76	67	76	64	75	59	75	53	79	55	82	55
	Dayton	12	75	67	76	63	78	62	75	56	77	54	79	54
OK	Oklahoma City	10	81	63	79	57	76	53	78	52	82	56	84	56
OR	Portland	35	82	76	79	68	72	60	68	55	66	53	65	49
PA	Philadelphia	16	74	60	71	57	71	53	69	49	75	53	78	55
	Pittsburgh	16	76	66	75	64	76	60	73	51	76	52	79	53
RI	Providence	12	71	56	70	55	69	54	69	48	72	52	76	57
SC	Columbia	9	85	58	82	50	83	48	84	45	88	51	89	54
SD	Sioux Falls	12	76	67	78	66	81	62	81	56	81	53	83	57
TN	Memphis	36	80	64	79	60	77	57	79	54	82	55	83	56
	Nashville	10	81	66	80	60	79	54	81	53	87	56	88	55
TX	Dallas-Fort Worth	12	82	61	80	57	80	57	85	59	88	62	86	57
	El Paso	15	43	33	35	25	29	21	21	15	21	14	24	17
	Houston	6	89	67	87	57	89	61	90	60	93	61	92	59
UT	Salt Lake City	16	69	67	63	58	51	44	45	39	36	31	32	27
VT	Burlington	10	69	63	71	61	72	58	75	53	75	52	79	58
VA	Norfolk	27	76	60	75	57	73	54	74	51	78	56	80	58
	Richmond	41	81	57	79	52	78	48	75	45	79	50	82	53
WA	Seattle-Tacoma	16	79	75	76	66	74	62	72	59	68	54	67	54
	Spokane	16	81	76	78	68	67	54	56	44	51	40	48	35
WV	Charleston	28	77	62	77	59	75	54	75	48	82	50	86	54
WI	Milwaukee	15	75	68	75	67	80	67	79	63	79	61	82	62
WY	Cheyenne	16	44	48	44	46	45	46	41	39	39	40	40	41
PR	San Juan	20	81	64	80	62	78	60	75	62	77	65	78	66

Source: U.S. National Oceanic and Atmospheric Administration, *Comparative Climatic Data.*
* City office data.

TABLE 3–4 Average Relative Humidity, Percent: Selected Cities
(continued)

July 7:00 a.m.	July 1:00 p.m.	Aug. 7:00 a.m.	Aug. 1:00 p.m.	Sept. 7:00 a.m.	Sept. 1:00 p.m.	Oct. 7:00 a.m.	Oct. 1:00 p.m.	Nov. 7:00 a.m.	Nov. 1:00 p.m.	Dec. 7:00 a.m.	Dec. 1:00 p.m.	Annual 7:00 a.m.	Annual 7:00 p.m.		State and Station
89	61	90	62	88	60	85	52	85	56	84	63	86	58	AL	Mobile
81	70	84	74	87	77	87	80	85	82	82	81	81	73	AK	Juneau
28	20	34	24	31	23	30	22	38	28	48	35	32	22	AZ	Phoenix
88	58	88	57	90	60	86	52	83	58	82	63	84	57	AR	Little Rock
68	68	68	69	65	67	58	64	58	64	55	61	62	65	CA	Los Angeles
48	28	49	28	50	31	58	40	77	61	86	73	63	46		Sacramento
66	61	67	62	66	59	68	59	74	65	78	69	70	62		San Francisco
36	36	36	34	40	36	36	36	45	50	45	52	40	40	CO	Denver
80	51	84	53	87	55	84	51	80	58	79	63	78	53	CN	Hartford
80	54	84	56	85	55	84	53	80	56	77	60	79	55	DE	Wilmington
75	52	78	53	79	55	79	51	73	52	71	57	73	52	DC	Washington
87	58	90	60	91	62	90	58	89	55	88	58	87	55	FL	Jacksonville
86	65	87	66	89	67	87	65	84	60	83	58	84	61		Miami
90	64	91	62	89	61	84	53	82	54	80	60	83	57	GA	Atlanta
65	51	67	54	66	52	68	55	74	60	78	61	71	57	HI	Honolulu
33	21	34	23	39	30	49	41	66	61	74	73	52	44	ID	Boise
77	55	80	56	81	56	78	54	79	65	80	72	76	59	IL	Chicago
86	59	88	59	88	60	85	58	83	67	84	73	83	62		Peoria
87	60	90	61	90	59	87	57	85	67	83	73	83	62	ID	Indianapolis
82	57	85	58	85	61	78	55	80	65	81	72	81	62	IO	Des Moines
79	49	79	49	82	56	81	53	79	57	80	63	80	55	KS	Wichita
86	58	87	57	89	60	85	55	79	61	77	66	81	59	KY	Louisville
91	66	91	66	89	66	88	59	86	60	86	67	88	63	LA	New Orleans
81	60	84	59	86	61	85	59	84	64	80	63	80	60	ME	Portland
81	53	84	55	85	55	82	53	77	54	75	59	77	54	MD	Baltimore
73	56	74	56	79	61	75	57	74	61	72	62	72	58	MA	Boston
75	51	80	53	83	54	81	55	79	64	79	70	77	58	MI	Detroit
88	62	91	63	92	67	89	67	87	76	85	78	85	67		Sault Ste. Marie
84	58	88	63	87	64	82	63	81	71	80	75	80	64	MN	Duluth
82	55	85	56	87	60	83	59	82	67	79	71	80	61		Minneapolis-St. Paul
93	59	94	60	94	60	93	54	91	58	89	66	90	59	MS	Jackson
77	48	85	58	89	66	81	58	86	65	79	70	80	60	MO	Kansas City
86	57	89	57	91	61	84	55	84	62	85	70	84	59		St. Louis
37	28	37	29	45	37	46	42	56	54	62	61	50	45	MT	Great Falls
82	56	86	59	88	61	82	55	81	63	80	68	80	59	NE	Omaha
29	19	31	20	34	21	41	27	57	42	66	53	44	32	NV	Reno
89	54	90	54	93	59	89	54	85	63	81	66	83	56	NH	Concord
85	59	87	57	88	58	87	56	82	57	77	59	81	57	NJ	Atlantic City
35	28	39	30	40	31	37	29	42	35	51	43	37	28	NM	Albuquerque
79	54	83	54	87	58	84	55	82	63	82	68	79	57	NY	Albany
79	55	83	59	83	60	82	61	83	71	82	75	80	63		Buffalo
75	55	78	57	79	57	76	55	73	59	70	61	72	56		New York*
89	60	90	60	90	57	88	53	83	51	80	57	84	55	NC	Charlotte
91	61	93	62	93	60	90	54	83	49	80	55	84	54		Raleigh
83	47	82	44	82	49	78	50	80	63	76	69	79	56	ND	Bismarck
82	57	85	60	84	61	80	59	78	67	76	71	79	63	OH	Cleveland
84	55	88	57	88	58	82	55	82	65	79	70	80	60		Columbus
80	54	85	56	86	58	81	56	80	66	79	71	79	60		Dayton
81	50	86	51	86	58	81	53	78	55	80	60	81	55	OK	Oklahoma City
61	45	64	46	66	49	79	64	82	74	84	79	72	60	OR	Portland
79	54	81	54	83	56	82	53	77	55	74	61	76	55	PA	Philadelphia
83	52	86	55	86	57	80	53	79	63	77	68	79	58		Pittsburgh
78	57	78	54	82	57	79	53	77	59	76	61	75	55	RI	Providence
91	58	93	59	94	58	92	52	88	49	86	56	88	53	SC	Columbia
82	52	84	52	87	58	81	56	84	65	80	71	81	60	SD	Sioux Falls
85	57	86	56	86	56	84	51	80	55	79	62	82	57	TN	Memphis
91	59	92	61	91	61	87	55	81	60	81	66	85	59		Nashville
81	50	83	53	88	60	85	56	82	56	81	60	83	57	TX	Dallas-Fort Worth
39	30	41	32	45	34	35	27	37	31	43	36	34	26		El Paso
93	58	95	62	95	66	95	60	90	59	88	62	91	61		Houston
26	20	28	22	34	27	43	40	58	59	71	72	46	42	UT	Salt Lake City
80	54	84	58	88	64	82	62	82	71	78	72	78	60	VT	Burlington
82	60	85	62	84	62	83	61	78	55	76	59	79	58	VA	Norfolk
85	57	88	57	89	56	89	52	84	50	81	55	83	53		Richmond
66	49	70	52	74	58	80	69	81	75	81	78	74	62	WA	Seattle-Tacoma
38	24	42	27	49	33	66	50	82	75	84	81	62	50		Spokane
90	61	92	58	91	55	88	53	80	56	78	62	83	56	WV	Charleston
82	60	87	62	87	63	81	63	80	67	80	73	81	65	WS	Milwaukee
34	37	33	34	37	38	37	39	41	47	44	50	40	42	WY	Cheyenne
79	66	79	66	80	67	81	66	81	66	80	65	79	64	PR	San Juan

TABLE 3–5 Normal Daily Mean Temperature: Selected Cities

(In Fahrenheit degrees. Airport data except New York City (city office). Based on standard 30-year period, 1941 through 1970.)

	State and Station	Jan.	Feb.	Mar.	Apr.	May	June	July	Aug.	Sept.	Oct.	Nov.	Dec.	Annual Average
AL	Mobile	51.2	54.0	59.4	67.9	74.8	80.3	81.6	81.5	77.5	68.9	58.5	52.9	67.4
AK	Juneau	23.5	28.0	31.9	38.9	46.8	53.2	55.7	54.3	49.2	41.8	32.5	27.3	40.3
AZ	Phoenix	51.2	55.1	59.7	67.7	76.3	84.6	91.2	89.1	83.8	72.2	59.8	52.5	70.3
AR	Little Rock	39.5	42.9	50.3	61.7	69.8	78.1	81.4	80.6	73.3	62.4	50.3	41.6	61.0
CA	Los Angeles	54.5	55.6	56.5	58.8	61.9	64.5	68.5	69.5	68.7	65.2	60.5	56.9	61.7
	Sacramento	45.1	49.8	53.0	58.3	64.3	70.5	75.2	74.1	71.5	63.3	53.0	45.8	60.3
	San Francisco	48.3	51.2	53.0	55.3	58.3	61.6	62.5	63.0	64.1	61.0	55.3	49.7	56.9
CO	Denver	29.9	32.8	37.0	47.5	57.0	66.0	73.0	71.6	62.8	52.0	39.4	32.6	50.1
CN	Hartford	24.8	26.8	35.6	47.7	58.3	67.8	72.7	70.4	62.8	52.6	41.3	28.2	49.1
DE	Wilmington	32.0	33.6	41.6	52.3	62.4	71.4	75.8	74.1	67.9	57.2	45.7	34.7	54.0
DC	Washington	35.6	37.3	45.1	56.4	66.2	74.6	78.7	77.1	70.6	59.8	48.0	37.4	57.3
FL	Jacksonville	54.6	56.3	61.2	68.1	74.3	79.2	81.0	81.0	78.2	70.5	61.2	55.4	68.4
	Miami	67.2	67.8	71.3	75.0	78.0	81.0	82.3	82.9	81.7	77.8	72.2	68.3	75.5
GA	Atlanta	42.4	45.0	51.1	61.1	69.1	75.6	78.0	77.5	72.3	62.4	51.4	43.5	60.8
HI	Honolulu	72.3	72.3	73.0	74.8	76.9	78.9	80.1	80.7	80.4	78.9	76.5	73.7	76.6
ID	Boise	29.0	35.5	41.1	49.0	57.4	64.8	74.5	72.2	63.1	52.1	39.8	32.1	50.9
IL	Chicago	24.3	27.4	36.8	49.9	60.0	70.5	74.7	73.7	65.9	55.4	40.4	28.5	50.6
	Peoria	23.8	27.7	37.3	51.3	61.5	71.3	75.1	73.5	65.5	55.0	39.9	28.0	50.8
IN	Indianapolis	27.9	30.7	39.7	52.3	62.2	71.7	75.0	73.2	66.3	55.7	41.7	30.9	52.3
IO	Des Moines	19.4	24.2	33.9	49.5	60.9	70.5	75.1	73.3	64.3	54.3	37.8	25.0	49.0
KS	Wichita	31.3	36.3	43.6	56.6	66.1	75.8	80.7	79.7	70.6	59.6	44.8	34.5	56.6
KY	Louisville	33.3	35.8	44.0	55.9	64.8	73.3	76.9	75.9	69.1	58.1	45.0	35.6	55.6
LA	New Orleans	52.9	55.6	60.7	68.6	75.1	80.4	81.9	81.9	78.2	69.8	60.1	54.8	68.3
ME	Portland	21.5	22.9	31.8	42.7	52.7	62.2	68.0	66.4	58.7	49.1	38.6	25.7	45.0
MD	Baltimore	33.4	34.8	42.8	53.8	63.7	72.4	76.6	74.9	68.5	57.4	46.1	35.3	55.0
MA	Boston	29.2	30.4	38.1	48.6	58.6	68.0	73.3	71.3	64.5	55.4	45.2	33.0	51.3
MI	Detroit	25.5	26.9	35.4	48.1	58.4	69.1	73.3	71.9	64.5	54.3	41.1	29.6	49.9
	Sault Ste. Marie	14.2	15.2	24.0	38.2	49.0	58.7	63.8	63.2	55.3	46.2	32.8	20.1	40.0
MN	Duluth	8.5	12.1	23.5	38.6	49.4	59.0	65.6	64.1	54.4	45.3	28.4	14.4	38.6
	Minneapolis-St. Paul	12.2	16.5	28.3	45.1	57.1	66.9	71.9	70.2	60.0	50.0	32.4	18.6	44.1
MS	Jackson	47.1	49.8	56.1	65.7	72.7	79.4	81.7	81.2	76.0	65.8	55.3	48.9	65.0
MO	Kansas City	27.8	33.1	41.2	55.0	65.0	73.9	78.8	77.4	68.8	58.6	43.6	32.3	54.5
	St. Louis	31.3	35.1	43.3	56.5	65.8	74.9	78.6	77.2	69.6	59.1	45.0	34.6	55.9
MT	Great Falls	20.5	26.6	30.5	43.4	53.3	60.8	69.3	67.4	57.3	48.3	34.6	26.5	44.9
NE	Omaha	22.6	28.0	37.1	52.3	63.0	72.2	77.2	75.6	66.3	55.9	40.0	28.0	51.5
NV	Reno	31.9	37.1	40.3	46.8	54.6	61.5	69.3	66.9	60.2	50.3	40.1	33.0	49.4
NH	Concord	20.6	22.6	32.3	44.2	55.1	64.7	69.7	67.2	59.5	49.3	38.0	24.8	45.6
NJ	Atlantic City	32.7	33.9	41.1	51.7	61.6	70.3	75.1	73.4	67.1	56.7	46.0	35.1	53.7
NM	Albuquerque	35.2	40.0	45.8	55.8	65.3	74.6	78.7	76.6	70.1	58.2	44.5	36.2	56.8
NY	Albany	21.5	23.5	33.4	46.9	57.7	67.5	72.0	69.6	61.9	51.4	39.6	25.9	47.6
	Buffalo	23.7	24.4	32.1	44.9	55.1	65.7	70.1	68.4	61.6	51.5	39.8	27.9	47.1
	New York	32.2	33.4	41.1	52.1	62.3	71.6	76.6	74.9	68.4	58.7	47.4	35.5	54.5
NC	Charlotte	42.1	44.0	50.6	60.8	68.8	75.9	78.5	77.7	72.0	61.7	51.0	42.5	60.5
	Raleigh	40.5	42.2	49.2	59.5	67.4	74.4	77.5	76.5	70.6	60.2	50.0	41.2	59.1
ND	Bismarck	8.2	13.5	25.1	43.0	54.4	63.8	70.8	69.2	57.5	46.8	28.9	15.6	41.4
OH	Cincinnati	32.1	34.4	42.9	55.1	64.4	73.1	76.2	75.1	68.4	57.8	44.6	34.4	54.9
	Cleveland	26.9	27.9	36.1	48.3	58.3	67.9	71.4	70.0	63.9	53.8	41.6	30.3	49.7
	Columbus	28.4	30.3	39.2	51.2	61.1	70.4	73.6	71.9	65.2	54.2	41.7	30.7	51.5
OK	Oklahoma City	36.8	41.3	48.2	60.4	68.3	76.8	81.5	81.1	73.0	62.4	49.2	40.0	59.9
OR	Portland	38.1	42.8	45.7	50.6	56.7	62.0	67.1	66.6	62.2	53.8	45.2	40.7	52.6
PA	Philadelphia	32.3	33.9	41.9	52.9	63.2	72.3	76.8	74.8	68.1	57.4	46.2	35.2	54.6
	Pittsburgh	28.1	29.3	38.1	50.2	59.8	68.6	71.9	70.2	63.8	53.2	41.3	30.5	50.4
RI	Providence	28.4	29.4	36.9	47.3	56.9	66.4	72.1	70.4	63.4	53.7	43.3	31.5	50.0
SC	Columbia	45.4	47.6	54.2	64.1	72.1	78.8	81.2	80.2	74.5	64.2	53.8	46.0	63.5
SD	Sioux Falls	14.2	19.4	30.0	46.1	57.7	67.6	73.3	71.8	60.9	50.2	33.1	20.0	45.4
TN	Memphis	40.5	43.8	51.0	62.5	70.9	78.6	81.6	80.4	73.6	63.0	50.9	42.7	61.6
	Nashville	38.3	41.0	48.7	60.1	68.5	76.6	79.6	78.5	72.0	60.9	48.4	40.4	59.4
TX	Dallas-Fort Worth	44.8	48.7	55.0	65.2	72.5	80.6	84.8	84.9	77.7	67.6	55.8	47.9	65.5
	El Paso	43.6	48.4	54.6	63.9	72.2	80.3	82.3	80.5	74.2	64.0	51.6	44.4	63.4
	Houston	52.1	55.3	60.8	69.4	75.8	81.1	83.3	83.4	79.2	70.9	61.1	54.6	68.9
UT	Salt Lake City	28.0	33.4	39.6	49.2	58.3	66.2	76.7	74.5	64.8	52.4	39.1	30.3	51.0
VT	Burlington	16.8	18.6	29.1	43.0	54.8	65.2	69.8	67.4	59.3	48.8	37.0	22.6	44.4
VA	Norfolk	40.5	41.4	48.1	57.8	66.7	74.5	78.3	76.9	71.8	61.7	51.6	42.3	59.3
	Richmond	37.5	39.4	46.9	57.8	66.5	74.2	77.9	76.3	70.0	59.3	49.0	39.0	57.8
WA	Seattle-Tacoma	38.2	42.4	44.1	48.7	54.9	59.8	64.5	63.8	59.6	52.2	44.6	40.5	51.1
	Spokane	25.4	32.2	37.5	46.1	54.7	61.5	69.7	68.0	59.6	47.8	35.5	29.0	47.3
WV	Charleston	34.5	36.5	44.5	55.9	64.5	72.0	75.0	73.6	67.5	57.0	45.4	36.2	55.2
WI	Milwaukee	19.4	22.5	31.4	44.7	54.2	64.5	69.9	69.2	61.1	51.0	36.5	24.2	45.7
WY	Cheyenne	26.6	29.0	31.6	42.7	52.4	61.3	69.1	67.6	58.2	47.9	35.5	29.2	45.9
PR	San Juan	75.4	75.3	76.3	77.5	79.2	80.5	80.9	81.3	81.1	80.6	78.7	76.8	78.6

Source: U.S. National Oceanic and Atmospheric Administration, *Comparative Climatic Data.*

TABLE 3–6 Normal Daily Minimum Temperature: Selected Cities

(In Fahrenheit degrees. Airport data unless otherwise noted. Based on standard 30-year period, 1941 through 1970.)

	State and Station	Jan.	Feb.	Mar.	Apr.	May	June	July	Aug.	Sept.	Oct.	Nov.	Dec.	Annual Average
AL	Mobile	41.3	43.9	49.2	57.7	64.5	70.7	72.6	72.3	68.4	58.0	47.5	42.8	57.4
AK	Juneau	17.8	22.1	25.6	31.3	38.2	44.4	47.7	46.2	42.3	36.4	27.6	22.5	33.5
AZ	Phoenix	37.6	40.8	44.8	51.8	59.6	67.7	77.5	76.0	69.1	56.8	44.8	38.5	55.4
AR	Little Rock	28.9	31.9	38.7	49.9	58.1	66.8	70.1	68.6	60.8	48.7	38.1	31.1	49.3
CA	Los Angeles	45.4	47.0	48.6	51.7	55.3	58.6	62.1	63.2	61.6	57.5	51.3	47.3	54.1
	Sacramento	37.1	40.4	41.9	45.3	49.8	54.6	57.5	56.9	55.3	49.5	42.4	38.3	47.4
	San Francisco	41.2	43.8	44.9	47.0	49.9	53.0	54.0	54.3	54.5	51.6	47.2	42.9	48.7
CO	Denver	16.2	19.4	23.8	33.9	43.6	51.9	58.6	57.4	47.8	37.2	25.4	18.9	36.2
CN	Hartford	16.1	17.9	26.6	36.5	46.2	56.0	61.2	58.9	51.0	40.8	31.9	19.6	38.6
DE	Wilmington	23.8	24.9	32.0	41.5	51.6	61.1	66.1	64.3	57.6	46.5	36.2	26.3	44.3
DC	Washington	27.7	28.6	35.2	45.7	55.7	64.6	69.1	67.6	61.0	49.7	38.8	29.5	47.8
FL	Jacksonville	44.5	45.7	50.1	57.1	63.9	70.0	72.0	72.3	70.4	61.7	51.0	45.1	58.7
	Miami	58.7	59.0	63.0	67.3	70.7	73.9	75.5	75.8	75.0	71.0	64.5	60.0	67.9
GA	Atlanta	33.4	35.5	41.1	50.7	59.2	66.6	69.4	68.6	63.4	52.3	40.8	34.3	51.3
HI	Honolulu	65.3	65.3	66.3	68.1	70.2	72.2	73.4	74.0	72.0	69.8	67.1	65.9	69.8
ID	Boise	21.4	27.2	30.5	36.5	44.1	51.2	58.5	56.7	48.5	39.4	30.7	25.0	39.1
IL	Chicago	17.0	20.2	29.0	40.4	49.7	60.3	65.0	64.1	56.0	45.6	32.6	21.6	41.8
	Peoria	15.7	19.3	28.1	40.8	50.7	60.9	64.6	62.9	54.6	44.0	31.1	20.3	41.1
IN	Indianapolis	19.7	22.1	30.3	41.8	51.5	61.1	64.6	62.4	54.9	44.3	32.8	23.1	42.4
IO	Des Moines	11.3	15.8	25.2	39.2	50.9	61.1	65.3	63.4	54.0	43.6	29.2	17.2	39.7
KS	Wichita	21.2	25.4	32.1	45.1	55.0	65.0	69.6	68.3	59.2	47.9	33.8	24.6	45.6
KY	Louisville	24.5	26.5	34.0	44.8	53.9	62.9	66.4	64.9	57.7	45.9	35.1	27.1	45.3
LA	New Orleans	43.5	46.0	50.9	58.8	65.3	71.2	73.3	73.1	69.7	59.6	49.8	45.3	58.9
ME	Portland	11.7	12.5	22.8	32.5	41.7	51.1	56.9	55.2	47.4	38.0	29.7	16.4	34.7
MD	Baltimore	24.9	25.7	32.5	42.4	52.5	61.6	66.5	64.7	57.9	46.4	36.0	26.6	44.8
MA	Boston	22.5	23.1	31.5	40.8	50.1	59.3	65.1	63.3	56.7	47.5	38.7	26.6	43.8
MI	Detroit	19.2	20.1	27.6	38.6	48.3	59.1	63.4	62.1	54.8	45.2	34.4	23.8	41.4
	Sault Ste. Marie	6.4	6.7	15.5	29.2	38.5	47.3	52.5	52.9	46.1	37.6	26.5	13.3	31.0
MN	Duluth	−.6	2.0	14.4	29.3	38.8	48.3	54.7	53.7	44.8	36.2	21.4	6.3	29.1
	Minneapolis-St. Paul	3.2	7.1	19.6	34.7	46.3	56.7	61.4	59.6	49.3	39.2	24.2	10.6	34.3
MS	Jackson	35.8	37.8	43.4	53.1	60.4	67.7	70.6	69.8	64.0	51.5	42.0	37.3	52.8
MO	Kansas City	19.3	24.2	31.8	45.1	55.7	65.2	69.6	68.1	58.8	48.3	34.5	24.1	45.3
	St. Louis	22.6	26.0	33.5	46.0	55.5	64.8	68.8	67.1	59.1	48.4	35.9	26.5	46.2
MT	Great Falls	11.6	17.2	20.6	32.3	41.5	49.5	54.9	53.0	44.6	37.1	25.7	18.2	33.8
NE	Omaha	12.4	17.4	26.4	40.1	51.5	61.3	65.8	64.0	54.0	42.6	29.1	18.1	40.2
NV	Reno	18.3	23.0	24.6	29.6	37.0	42.5	47.4	44.8	38.6	30.5	23.9	19.6	31.7
NH	Concord	9.9	11.3	22.1	31.7	41.5	51.6	56.7	54.2	46.5	36.3	28.1	14.9	33.7
NJ	Atlantic City	24.0	24.9	31.5	41.0	50.7	59.7	65.4	63.8	56.8	45.9	36.1	26.0	43.8
NM	Albuquerque	23.5	27.4	32.3	41.4	50.7	59.7	65.2	63.4	56.7	44.7	31.8	24.9	43.5
NY	Albany	12.5	14.3	24.2	35.7	45.7	55.6	60.1	57.8	50.1	40.0	31.1	17.7	37.1
	Buffalo	17.6	17.7	25.2	36.4	45.9	56.3	60.7	59.1	52.3	42.7	33.5	22.2	39.1
	New York*	25.9	26.5	33.7	43.5	53.1	62.6	68.0	66.4	59.9	50.6	40.8	29.5	46.7
NC	Charlotte	32.1	33.1	39.0	48.9	57.4	65.3	68.7	67.9	61.9	50.3	39.6	32.4	49.7
	Raleigh	30.0	31.1	37.4	46.7	55.4	63.1	67.2	66.2	59.7	48.0	37.8	30.5	47.8
ND	Bismarck	−2.8	2.4	14.7	31.1	41.7	51.8	57.3	54.9	43.7	33.2	18.3	5.2	29.3
OH	Cincinnati	24.3	25.8	33.5	44.6	53.6	62.5	65.8	64.1	57.0	46.7	36.2	27.1	45.1
	Cleveland	20.3	20.8	28.1	38.5	48.1	57.5	61.2	59.6	53.5	43.9	34.4	24.1	40.8
	Columbus	20.4	21.4	29.1	39.5	49.3	58.9	62.4	60.1	52.7	42.0	32.4	22.7	40.9
OK	Oklahoma City	26.0	30.0	36.5	49.1	57.9	66.6	70.4	69.6	61.3	50.6	37.4	29.2	48.7
OR	Portland	32.5	35.5	37.0	40.8	46.3	51.8	55.2	55.0	50.5	44.7	38.5	35.3	43.6
PA	Philadelphia	24.4	25.5	32.5	42.3	52.3	61.6	66.6	64.7	57.8	46.9	36.9	27.2	44.9
	Pittsburgh	20.8	21.3	29.0	39.4	48.7	57.7	61.3	59.4	52.7	42.4	33.3	23.6	40.8
RI	Providence	20.6	21.2	29.0	37.8	46.9	56.5	63.0	61.0	53.6	43.4	34.6	23.4	40.9
SC	Columbia	33.9	35.5	41.9	51.3	59.6	67.2	70.3	69.4	63.5	51.3	40.6	34.1	51.5
SD	Sioux Falls	3.7	9.0	20.2	34.4	45.7	56.3	61.5	59.8	48.7	37.6	22.7	10.4	34.2
TN	Memphis	31.6	34.4	41.1	52.3	60.6	68.5	71.5	70.1	62.8	51.1	40.3	33.7	51.5
	Nashville	29.0	31.0	38.1	48.8	57.3	65.7	69.0	67.7	60.5	48.6	37.7	31.1	48.7
TX	Dallas-Fort Worth	33.9	37.6	43.3	54.1	62.1	70.3	74.0	73.7	66.8	56.0	44.1	37.0	54.4
	El Paso	30.2	34.3	40.3	49.3	57.2	65.7	69.9	68.2	61.0	49.5	37.0	30.9	49.5
	Houston	41.5	44.6	49.8	59.3	65.6	70.9	72.4	72.2	68.2	58.3	49.1	43.4	58.0
UT	Salt Lake City	18.5	23.3	28.3	36.6	44.2	51.1	60.5	58.7	49.3	38.4	28.1	21.5	38.2
VT	Burlington	7.6	8.9	20.1	32.6	43.5	53.9	58.5	56.4	48.6	38.8	29.7	14.8	34.5
VA	Norfolk	32.2	32.7	38.9	47.9	57.2	65.7	69.9	68.9	63.9	53.3	42.6	34.0	50.6
	Richmond	27.6	28.8	35.5	45.2	54.5	62.9	67.5	65.9	59.0	47.4	37.3	28.8	46.7
WA	Seattle-Tacoma	33.0	36.0	36.6	40.3	45.6	50.6	53.8	53.7	50.4	44.9	38.8	35.5	43.3
	Spokane	19.6	25.3	28.8	35.2	42.8	49.4	55.1	54.0	46.7	37.5	29.2	24.0	37.3
WV	Charleston	25.3	26.8	33.8	43.8	52.3	60.6	64.3	62.8	55.9	44.8	35.0	27.2	44.4
WI	Milwaukee	11.4	14.6	23.4	34.7	43.3	53.6	59.3	58.7	50.7	40.6	28.5	16.8	36.3
WY	Cheyenne	14.9	17.3	19.6	30.0	39.7	48.1	54.5	53.2	43.5	33.9	23.5	18.1	33.0
PR	San Juan	68.8	68.4	68.9	70.6	72.8	74.0	74.8	75.1	74.6	73.7	72.3	70.5	72.0

Source: U.S. National Oceanic and Atmospheric Administration, *Comparative Climatic Data.*
* City office data.

TABLE 3–7 Normal Daily Maximum Temperature: Selected Cities

(In Fahrenheit degrees. Airport data unless otherwise noted. Based on standard 30-year period, 1941 through 1970.)

	State and Station	Jan.	Feb.	Mar.	Apr.	May	June	July	Aug.	Sept.	Oct.	Nov.	Dec.	Annual Average
AL	Mobile	61.1	64.1	69.5	78.0	85.0	89.8	90.5	90.6	86.5	79.7	69.5	63.0	77.3
AK	Juneau	29.1	33.9	38.2	46.5	55.4	62.0	63.6	62.3	56.1	47.2	37.3	32.0	47.0
AZ	Phoenix	64.8	69.3	74.5	83.6	92.9	101.5	104.8	102.2	98.4	87.6	74.7	66.4	85.1
AR	Little Rock	50.1	53.8	61.8	73.5	81.4	89.3	92.6	92.6	85.8	76.0	62.4	52.1	72.6
CA	Los Angeles	63.5	64.1	64.3	65.9	68.4	70.3	74.8	75.8	75.7	72.9	69.6	66.5	69.2
	Sacramento	53.0	59.1	64.1	71.3	78.8	86.4	92.9	91.3	87.7	77.1	63.6	53.3	73.2
	San Francisco	55.3	58.6	61.0	63.5	66.6	70.2	70.9	71.6	73.6	70.3	63.3	56.5	65.1
CO	Denver	43.5	46.2	50.1	61.0	70.3	80.1	87.4	85.8	77.7	66.8	53.3	46.2	64.0
CN	Hartford	33.4	35.7	44.6	58.9	70.3	79.5	84.1	81.9	74.5	64.3	50.6	36.8	59.6
DE	Wilmington	40.2	42.2	51.1	63.0	73.1	81.6	85.5	83.9	78.2	67.8	55.2	43.0	63.7
DC	Washington	43.5	46.0	55.0	67.1	76.6	84.6	88.2	86.6	80.2	69.8	57.2	45.2	66.7
FL	Jacksonville	64.6	66.9	72.2	79.0	84.6	88.3	90.0	89.7	86.0	79.2	71.4	65.6	78.1
	Miami	75.6	76.6	79.5	82.7	85.3	88.0	89.1	89.9	88.3	84.6	79.9	76.6	83.0
GA	Atlanta	51.4	54.5	61.1	71.4	79.0	84.6	86.5	86.4	81.2	72.5	61.9	52.7	70.3
HI	Honolulu	79.3	79.2	79.7	81.4	83.6	85.6	86.8	87.4	87.4	85.8	83.2	80.3	83.3
ID	Boise	36.5	43.8	51.6	61.4	70.6	78.3	90.5	87.6	77.6	64.7	48.9	39.1	62.6
IL	Chicago	31.5	34.6	44.6	59.3	70.3	80.6	84.4	83.3	75.8	65.1	48.1	35.3	59.4
	Peoria	31.9	36.0	46.5	61.7	72.3	81.7	85.5	84.0	76.4	65.9	48.7	35.7	60.5
IN	Indianapolis	36.0	39.3	49.0	62.8	72.9	82.3	85.4	84.0	77.7	67.0	50.5	38.7	62.2
IO	Des Moines	27.5	32.5	42.5	59.7	70.9	79.8	84.9	83.2	74.6	64.9	46.4	32.8	58.3
KS	Wichita	41.4	47.1	55.0	68.1	77.1	86.5	91.7	91.0	81.9	71.3	55.8	44.3	67.6
KY	Louisville	42.0	45.0	54.0	66.9	75.6	83.7	87.3	86.8	80.5	70.3	54.9	44.1	65.9
LA	New Orleans	62.3	65.1	70.4	78.4	84.9	89.6	90.4	90.6	86.6	79.9	70.3	64.2	77.7
ME	Portland	31.2	33.3	40.8	52.8	63.6	73.2	79.1	77.6	69.9	60.2	47.5	34.9	55.3
MD	Baltimore	41.9	43.9	53.0	65.2	74.8	83.2	86.7	85.1	79.0	68.3	56.1	43.9	65.1
MA	Boston	35.9	37.5	44.6	56.3	67.1	76.6	81.4	79.3	72.2	63.2	51.7	39.3	58.7
MI	Detroit	31.7	33.7	43.1	57.6	68.5	79.1	83.1	81.6	74.2	63.4	47.7	35.4	58.3
	Sault Ste. Marie	22.0	23.7	32.5	47.2	59.4	70.0	75.1	73.4	64.5	54.8	39.0	26.8	49.0
MN	Duluth	17.6	22.1	32.6	47.8	60.0	69.7	76.4	74.4	64.3	54.3	35.3	22.5	48.1
	Minneapolis-St. Paul	21.2	25.9	36.9	55.5	67.9	77.1	82.4	80.8	70.7	60.7	40.6	26.6	53.8
MS	Jackson	58.4	61.7	68.7	78.2	85.0	91.0	92.7	92.6	88.0	80.1	68.5	60.5	77.1
MO	Kansas City	36.2	41.9	50.5	64.8	74.3	82.6	88.0	86.7	78.8	68.9	52.7	40.4	63.7
	St. Louis	39.9	44.2	53.0	67.0	76.0	84.9	88.4	87.2	80.1	69.8	54.1	42.7	65.6
MT	Great Falls	29.3	35.9	40.4	54.5	65.0	72.1	83.7	81.8	70.0	59.4	43.4	34.7	55.9
NE	Omaha	32.7	38.5	47.7	64.4	74.4	83.1	88.6	87.2	78.6	69.1	50.9	37.8	62.8
NV	Reno	45.4	51.1	56.0	64.0	72.2	80.4	91.1	89.0	81.8	70.0	56.3	46.4	67.0
NH	Concord	31.3	33.8	42.4	56.7	68.6	77.7	82.6	80.1	72.4	62.3	47.9	34.6	57.5
NJ	Atlantic City	41.4	42.9	50.7	62.3	72.4	80.8	84.7	83.0	77.6	67.5	55.9	44.2	63.6
NM	Albuquerque	46.9	52.6	59.2	70.1	79.9	89.5	92.2	89.7	83.4	71.7	57.1	47.5	70.0
NY	Albany	30.4	32.7	42.6	58.0	69.7	79.4	83.9	81.4	73.7	62.8	48.1	34.1	58.1
	Buffalo	29.8	31.0	39.0	53.3	64.3	75.1	79.5	77.6	70.8	60.2	46.1	33.6	55.0
	New York*	38.5	40.2	48.4	60.7	71.4	80.5	85.2	83.4	76.8	66.8	54.0	41.4	62.3
NC	Charlotte	52.1	54.9	62.2	72.7	80.2	86.4	88.3	87.4	82.0	73.1	62.4	52.5	71.2
	Raleigh	51.0	53.2	61.0	72.2	79.4	85.6	87.7	86.8	81.5	72.4	62.1	51.9	70.4
ND	Bismarck	19.1	24.5	35.4	54.8	67.1	75.8	84.3	83.5	71.3	60.3	39.4	26.0	53.5
OH	Cincinnati	39.8	42.9	52.2	65.5	75.2	83.6	86.6	86.0	79.8	68.8	53.0	41.8	64.6
	Cleveland	33.4	35.0	44.1	58.0	68.4	78.2	81.6	80.4	74.2	63.6	48.8	36.4	58.5
	Columbus	36.4	39.2	49.3	62.8	72.9	81.9	84.8	83.7	77.6	66.4	50.9	38.7	62.1
OK	Oklahoma City	47.6	52.6	59.8	71.6	78.7	87.0	92.6	92.5	84.7	74.2	60.9	50.7	71.1
OR	Portland	43.6	50.1	54.3	60.3	67.0	72.1	79.0	78.1	73.9	62.9	52.1	46.0	61.6
PA	Philadelphia	40.1	42.2	51.2	63.5	74.1	83.0	86.8	84.8	78.4	67.9	55.5	43.2	64.2
	Pittsburgh	35.3	37.3	47.2	60.9	70.8	79.5	82.5	80.9	74.9	63.9	49.3	37.3	60.0
RI	Providence	36.2	37.6	44.7	56.7	66.8	76.3	81.1	79.8	73.3	52.0	39.6		59.0
SC	Columbia	56.9	59.7	66.5	76.9	84.5	90.3	92.0	91.0	85.4	77.1	66.9	57.9	75.4
SD	Sioux Falls	24.6	29.7	39.7	57.8	69.7	78.9	85.1	83.8	73.0	62.7	43.5	29.6	56.5
TN	Memphis	49.4	53.1	60.8	72.7	81.2	88.7	91.6	90.6	84.3	74.9	61.5	51.7	71.7
	Nashville	47.6	50.9	59.2	71.3	79.8	87.5	90.2	89.2	83.5	73.2	59.0	49.6	70.1
TX	Dallas-Fort Worth	55.7	59.8	66.6	76.3	82.8	90.8	95.5	96.1	88.5	79.2	67.5	58.7	76.5
	El Paso	57.0	62.5	68.9	78.5	87.2	94.9	94.6	92.8	87.4	78.5	66.1	57.8	77.2
	Houston	62.6	66.0	71.8	79.4	85.9	91.3	93.8	94.3	90.1	83.5	73.0	65.8	79.8
UT	Salt Lake City	37.4	43.4	50.8	61.8	72.4	81.3	92.8	90.2	80.3	66.4	50.0	39.0	63.8
VT	Burlington	25.9	28.2	38.0	53.3	66.1	76.5	81.0	78.3	70.0	58.7	44.3	30.3	54.2
VA	Norfolk	48.8	50.0	57.3	67.7	76.2	83.5	86.6	84.9	79.6	70.1	60.5	50.6	68.0
	Richmond	47.4	49.9	58.2	70.3	78.4	85.4	88.2	86.6	80.9	71.2	60.6	49.1	68.8
WA	Seattle-Tacoma	43.4	48.5	51.5	57.0	64.1	69.0	75.1	73.8	68.7	59.4	50.4	45.4	58.8
	Spokane	31.1	39.0	46.2	57.0	66.5	73.6	84.3	81.9	72.8	58.1	41.8	33.9	57.2
WV	Charleston	43.6	46.2	55.2	67.9	76.6	83.4	85.6	84.4	79.0	69.1	55.8	45.2	66.0
WI	Milwaukee	27.3	30.3	39.4	54.6	65.0	75.3	80.4	79.7	71.5	61.4	44.4	31.5	55.1
WY	Cheyenne	38.2	40.7	43.5	55.4	65.1	74.4	83.7	81.9	72.8	61.8	47.5	40.3	58.8
PR	San Juan	81.9	82.1	83.6	84.4	85.6	87.0	87.0	87.5	87.6	87.4	85.0	83.1	85.2

Source: U.S. National Oceanic and Atmospheric Administration, *Comparative Climatic Data.*
* City office data.

TABLE 3-8 Highest Temperature of Record: Selected Cities

(In Fahrenheit degrees. Airport data unless otherwise noted. For period of record through 1975.)

	State and Station	Length of Record (Years)	Jan.	Feb.	Mar.	Apr.	May	June	July	Aug.	Sept.	Oct.	Nov.	Dec.	Annual
AL	Mobile	14	79	81	89	91	99	101	100	102	98	93	87	81	102
AK	Juneau	32	57	50	55	71	82	86	90	83	72	61	56	54	90
AZ	Phoenix	15	88	89	95	101	110	116	115	116	110	103	93	82	116
AR	Little Rock	16	81	83	91	90	98	102	105	108	102	97	85	79	108
CA	Los Angeles	17	87	92	88	95	96	92	92	91	110	106	101	88	110
	Sacramento	25	69	76	86	92	102	115	114	108	108	101	87	72	115
	San Francisco	16	71	72	79	85	94	106	98	98	103	95	85	72	106
CO	Denver	16	69	76	84	84	93	98	103	100	97	87	78	73	103
CN	Hartford	16	65	59	77	94	96	100	102	101	96	91	81	65	102
DE	Wilmington	28	75	74	86	91	95	99	102	101	100	91	85	72	102
DC	Washington	15	76	77	86	91	97	100	101	99	96	91	86	74	101
FL	Jacksonville	34	85	88	91	95	100	103	105	102	100	96	88	84	105
	Miami	11	86	88	90	96	93	94	96	96	93	90	87	85	96
GA	Atlanta	15	77	79	85	88	93	98	98	98	96	88	84	77	98
HI	Honolulu	6	85	85	87	87	88	90	90	91	92	91	89	85	92
ID	Boise	36	63	67	78	92	98	109	111	110	102	91	73	65	111
IL	Chicago	12	65	65	80	88	94	101	99	98	95	94	78	71	101
	Peoria	16	66	70	81	87	92	100	102	99	94	90	77	71	102
IN	Indianapolis	17	70	74	80	89	93	96	99	97	96	88	78	70	99
IO	Des Moines	15	62	73	83	90	98	99	104	100	95	95	76	67	104
KS	Wichita	23	75	82	89	96	100	106	113	110	103	95	80	83	113
KY	Louisville	15	73	77	83	88	91	97	101	101	96	89	82	73	101
LA	New Orleans	29	83	85	87	91	96	100	99	100	97	92	86	84	100
ME	Portland	35	64	64	86	85	92	97	98	103	95	88	74	62	103
MD	Baltimore	25	75	76	85	94	98	100	102	102	99	92	83	74	102
MA	Boston	11	63	58	70	85	93	97	98	102	95	86	77	70	102
MI	Detroit	41	67	68	82	87	93	104	105	101	100	92	81	66	105
	Sault Ste. Marie	35	45	45	75	82	89	92	97	98	93	80	66	59	98
MN	Duluth	15	47	44	65	83	90	92	94	95	90	84	69	55	95
	Minneapolis-St. Paul	16	46	54	83	91	95	99	101	98	95	87	74	62	101
MS	Jackson	12	82	82	88	92	99	103	102	99	98	91	88	81	103
MO	Kansas City	3	61	66	82	85	91	98	107	103	98	89	73	67	107
	St. Louis	15	76	85	88	92	92	98	106	105	100	94	81	76	106
MT	Great Falls	15	61	64	72	83	90	97	105	106	98	85	76	61	106
NE	Omaha	12	64	78	89	93	97	103	110	107	103	95	80	67	110
NV	Reno	12	70	74	83	88	95	100	103	103	96	91	76	70	103
NH	Concord	10	60	57	68	88	96	97	102	101	93	84	78	61	102
NJ	Atlantic City	11	78	70	81	94	99	106	104	97	93	87	81	72	106
NM	Albuquerque	16	69	75	85	89	95	105	104	99	95	87	77	68	105
NY	Albany	10	62	57	77	88	92	98	98	97	93	84	77	65	98
	Buffalo	15	61	61	78	83	88	94	94	93	90	82	80	66	94
	New York*	107	72	75	86	92	99	101	106	104	102	94	84	70	106
NC	Charlotte	15	77	78	86	91	95	99	99	100	94	87	85	77	100
	Raleigh	11	77	79	89	93	92	95	97	98	93	89	84	77	98
ND	Bismarck	16	54	61	80	91	95	100	109	107	100	95	75	62	109
OH	Cincinnati	60	77	77	88	90	95	102	109	103	101	92	83	71	109
	Cleveland	15	68	69	80	85	91	95	98	95	93	86	79	69	98
	Columbus	16	68	72	80	88	93	96	97	98	96	86	79	72	98
OK	Oklahoma City	10	79	81	93	100	96	102	108	106	101	96	84	80	108
OR	Portland	35	62	70	80	87	92	100	107	104	101	90	73	64	107
PA	Philadelphia	16	69	69	80	92	96	100	104	99	97	88	81	71	104
	Pittsburgh	16	68	66	80	87	91	96	98	96	95	87	82	72	98
RI	Providence	12	66	59	73	90	94	95	97	104	93	85	78	69	104
SC	Columbia	9	84	81	91	94	96	104	103	106	97	90	89	83	106
SD	Sioux Falls	12	57	59	87	92	100	101	106	108	101	94	76	60	108
TN	Memphis	34	78	81	85	91	97	104	106	105	103	95	85	79	106
	Nashville	10	78	79	86	88	91	98	103	99	95	90	84	74	103
TX	Dallas-Fort Worth	12	88	87	96	95	96	105	106	108	102	96	88	84	108
	El Paso	16	80	83	88	98	101	108	106	105	100	96	84	80	108
	Houston	6	84	82	90	89	93	99	101	101	97	93	88	83	101
UT	Salt Lake City	16	61	69	78	85	92	104	107	103	96	89	75	67	107
VT	Burlington	11	56	51	67	84	91	93	98	99	90	80	71	62	99
VA	Norfolk	27	78	79	85	97	97	101	103	99	98	95	86	79	103
	Richmond	46	80	83	93	96	100	104	104	102	103	99	86	80	104
WA	Seattle-Tacoma	16	61	70	71	77	93	94	97	99	93	81	72	60	99
	Spokane	16	59	60	71	80	92	100	103	108	93	85	67	53	108
WV	Charleston	28	79	77	87	91	93	98	102	100	102	92	85	80	102
WI	Milwaukee	16	57	51	77	85	92	95	98	99	94	89	74	63	99
WY	Cheyenne	16	62	71	73	82	90	94	98	96	93	83	70	66	98
PR	San Juan	21	90	92	93	93	94	96	93	96	94	95	92	90	96

Source: U.S. National Oceanic and Atmospheric Administration, *Comparative Climatic Data.*
* City office data.

TABLE 3–9 Lowest Temperature of Record: Selected Cities

(In Fahrenheit degrees. Airport data unless otherwise noted. For period of record through 1975.)

	State and Station	Length of Record (Years)	Jan.	Feb.	Mar.	Apr.	May	June	July	Aug.	Sept.	Oct.	Nov.	Dec.	Annual
AL	Mobile	14	8	11	11	36	46	56	62	60	42	38	24	10	8
AK	Juneau	32	−22	−22	−15	6	25	31	36	27	23	12	−5	−21	−22
AZ	Phoenix	15	19	26	25	37	40	51	67	61	47	34	31	24	19
AR	Little Rock	16	−4	10	17	28	40	46	54	52	38	31	17	−1	−4
CA	Los Angeles	17	30	37	39	43	45	50	55	58	55	43	38	32	30
	Sacramento	25	23	26	26	32	36	41	49	49	43	36	26	20	20
	San Francisco	16	29	35	31	38	40	45	48	49	45	39	35	24	24
CO	Denver	16	−25	−18	−4	−2	26	36	43	41	20	3	−2	−18	−25
CN	Hartford	16	−26	−21	−6	9	30	37	44	36	30	18	12	−9	−26
DE	Wilmington	28	−4	−4	9	22	32	41	50	46	36	24	14	3	−4
DC	Washington	15	3	4	16	27	36	47	56	51	39	29	20	10	3
FL	Jacksonville	34	19	19	25	35	45	56	61	64	50	38	21	12	12
	Miami	11	35	36	37	46	61	67	70	70	70	56	40	34	34
GA	Atlanta	15	−3	8	20	26	37	48	53	56	36	29	14	1	−3
HI	Honolulu	6	53	54	58	59	63	65	67	67	66	64	58	54	53
ID	Boise	36	−17	−10	6	19	26	33	41	37	23	11	−3	−23	−23
IL	Chicago	12	−16	−9	5	16	29	41	46	43	34	26	7	−10	−16
	Peoria	16	−20	−14	−10	17	25	40	47	44	31	19	4	−18	−20
IN	Indianapolis	17	−20	−10	−6	18	28	42	48	41	34	20	4	−14	−20
IO	Des Moines	15	−24	−18	−22	9	30	42	47	45	31	14	−3	−16	−24
KS	Wichita	23	−12	−6	−2	15	32	43	51	48	35	23	1	−5	−12
KY	Louisville	15	−20	−4	15	24	31	42	50	49	37	25	10	−3	−20
LA	New Orleans	29	14	19	26	32	41	55	60	60	42	35	24	17	14
ME	Portland	35	−26	−39	−21	8	23	33	40	33	23	18	5	−21	−39
MD	Baltimore	25	−7	−1	6	20	32	40	52	48	35	25	13	0	−7
MA	Boston	11	−4	−3	6	22	37	46	54	47	38	30	17	−3	−4
MI	Detroit	41	−13	−16	−1	14	30	38	42	43	32	24	5	−5	−16
	Sault Ste. Marie	35	−28	−28	−24	1	18	28	36	32	25	16	−5	−20	−28
MN	Duluth	15	−39	−32	−28	−5	17	27	36	33	23	12	−23	−29	−39
	Minneapolis-St. Paul	16	−34	−28	−32	2	18	37	43	39	26	15	−17	−24	−34
MS	Jackson	12	7	11	18	30	38	49	51	55	35	30	17	14	7
MO	Kansas City	3	−13	−4	−1	12	34	47	52	50	39	28	7	−7	−13
	St. Louis	15	−11	−2	3	22	31	43	51	47	36	25	1	−6	−11
MT	Great Falls	15	−37	−24	−19	−6	19	32	40	37	23	−3	−15	−43	−43
NE	Omaha	12	−22	−19	−1	5	31	40	44	43	31	13	−9	−13	−22
NV	Reno	12	−11	0	0	15	18	29	33	29	20	8	5	−16	−16
NH	Concord	10	−29	−27	−16	8	21	30	35	29	22	10	−1	−18	−29
NJ	Atlantic City	11	−8	−7	7	12	25	37	46	40	32	23	11	0	−8
NM	Albuquerque	16	−17	1	9	22	28	42	54	52	37	25	10	3	−17
NY	Albany	10	−28	−21	−10	10	26	36	43	37	28	16	5	−22	−28
	Buffalo	15	−11	−20	−4	13	29	36	46	38	32	20	9	−4	−20
	New York*	107	−6	−15	3	12	32	44	52	50	39	28	5	−13	−15
NC	Charlotte	15	4	7	18	25	32	45	53	53	39	24	13	2	2
	Raleigh	11	0	5	17	23	33	43	48	46	39	24	11	9	0
ND	Bismarck	16	−42	−37	−28	−12	15	30	35	33	11	5	−29	−43	−43
OH	Cincinnati	60	−17	−9	3	18	28	40	48	43	32	20	1	−13	−17
	Cleveland	15	−19	−15	4	10	25	31	41	41	34	22	13	−4	−19
	Columbus	16	−15	−11	−2	18	25	35	43	39	31	20	11	−10	−15
OK	Oklahoma City	10	−1	3	9	21	39	51	53	54	37	31	13	1	−1
OR	Portland	35	−2	−3	19	29	29	39	43	44	34	26	13	6	−3
PA	Philadelphia	16	−5	−4	9	24	28	44	51	45	35	25	17	3	−5
	Pittsburgh	16	−18	−9	−1	15	26	34	42	40	31	16	8	−5	−18
RI	Providence	12	−5	−5	1	19	32	41	49	40	34	21	14	−4	−5
SC	Columbia	9	5	5	18	29	36	46	59	53	40	25	12	15	5
SD	Sioux Falls	12	−36	−30	−14	5	17	33	38	37	22	9	−17	−26	−36
TN	Memphis	34	−4	−11	12	29	38	48	52	48	36	25	9	−13	−13
	Nashville	10	−6	−5	14	24	35	42	54	51	37	27	12	6	−6
TX	Dallas-Ft. Worth	12	4	12	19	30	42	51	59	56	46	37	22	10	4
	El Paso	16	−8	11	14	24	31	51	59	56	42	25	18	10	−8
	Houston	6	19	22	25	31	46	52	62	62	48	39	24	21	19
UT	Salt Lake City	16	−18	−4	2	22	25	35	40	37	27	16	11	−15	−18
VT	Burlington	11	−27	−25	−13	2	24	33	40	35	29	15	2	−23	−27
VA	Norfolk	27	8	8	20	28	36	45	56	52	45	29	20	14	8
	Richmond	46	−12	−10	11	26	31	40	51	46	35	21	10	−1	−12
WA	Seattle-Tacoma	16	12	18	23	29	33	41	45	45	35	30	22	6	6
	Spokane	16	−19	−12	1	17	26	34	38	35	25	13	−2	−25	−25
WV	Charleston	28	−12	−6	7	19	26	33	46	41	34	17	6	−2	−12
WI	Milwaukee	16	−24	−15	−10	13	21	36	40	44	28	21	6	−15	−24
WY	Cheyenne	16	−27	−24	−11	−8	21	34	39	36	22	7	−8	−24	−27
PR	San Juan	21	61	62	60	64	66	69	69	70	69	67	66	63	60

Source: U.S. National Oceanic and Atmospheric Administration, *Comparative Climatic Data.*
* City office data.

TABLE 3-10 Average Wind Speed: Selected Cities

(In miles per hour. Airport data, except as noted. For period of record through 1975.)

	State and Station	Length of Record (Years)	Jan.	Feb.	Mar.	Apr.	May	June	July	Aug.	Sept.	Oct.	Nov.	Dec.	Annual Average
AL	Mobile	27	10.8	11.0	11.3	10.7	9.1	7.9	7.1	6.9	8.2	8.4	9.6	10.3	9.3
AK	Juneau	32	8.5	8.8	8.8	8.9	8.4	7.9	7.6	7.6	8.0	9.7	8.7	9.4	8.5
AZ	Phoenix	30	5.1	5.7	6.4	6.8	6.8	6.8	7.1	6.5	6.2	5.7	5.2	5.0	6.1
AR	Little Rock	33	8.9	9.4	10.1	9.6	8.1	7.6	7.0	6.6	7.0	7.0	8.2	8.5	8.2
CA	Los Angeles	27	6.7	7.3	8.0	8.4	8.2	7.8	7.6	7.5	7.1	6.8	6.6	6.6	7.4
	Sacramento	27	8.0	8.0	9.0	9.1	9.4	10.0	9.2	8.7	7.9	6.9	6.5	7.2	8.3
	San Francisco	48	7.1	8.5	10.3	12.1	13.1	13.9	13.6	12.8	11.0	9.2	7.2	6.8	10.5
CO	Denver	27	9.3	9.3	10.0	10.4	9.5	9.1	8.6	8.2	8.2	8.2	8.7	9.0	9.0
CN	Hartford	21	9.6	9.9	10.5	10.7	9.5	8.5	7.9	7.7	7.8	8.2	8.8	9.0	9.0
DE	Wilmington	27	9.7	10.5	11.2	10.5	9.0	8.4	7.7	7.4	7.8	8.1	9.1	9.3	9.1
DC	Washington	27	9.9	10.4	10.9	10.5	9.2	8.7	8.1	8.0	8.2	8.5	9.2	9.4	9.2
FL	Jacksonville	26	8.4	9.6	9.6	9.3	8.8	8.5	7.6	7.4	8.4	8.8	8.3	8.2	8.6
	Miami	26	9.3	10.0	10.3	10.6	9.4	8.2	7.8	7.7	8.2	9.3	9.4	9.0	9.1
GA	Atlanta	37	10.5	11.0	11.0	10.1	8.6	7.9	7.4	7.1	8.0	8.3	9.1	9.8	9.1
HI	Honolulu	26	10.0	10.8	11.4	12.1	12.2	12.9	13.6	13.6	11.7	10.9	11.2	11.1	11.8
ID	Boise	36	8.6	9.4	10.4	10.3	9.6	9.2	8.5	8.3	8.3	8.6	8.7	8.5	9.0
IL	Chicago	33	11.4	11.6	11.9	11.8	10.4	9.4	8.3	8.1	9.0	9.8	11.3	11.1	10.3
	Peoria	32	11.2	11.6	12.3	12.3	10.4	9.1	8.0	7.8	8.8	9.5	11.2	10.9	10.3
IN	Indianapolis	27	11.1	11.2	11.9	11.5	9.7	8.5	7.4	7.2	8.1	8.9	10.7	10.5	8.7
IO	Des Moines	26	11.7	11.8	13.1	13.4	11.6	10.5	9.0	8.8	9.6	10.6	11.7	11.5	11.1
KS	Wichita	22	12.5	13.1	14.5	14.7	13.1	12.6	11.3	11.3	11.7	12.3	12.4	12.2	12.6
KY	Louisville	28	9.6	9.8	10.4	10.1	8.1	7.4	6.7	6.4	6.8	7.1	9.0	9.3	8.4
LA	New Orleans	27	9.5	10.1	10.3	9.8	8.3	7.0	6.3	6.1	7.5	7.6	8.9	9.2	8.4
ME	Portland	35	9.2	9.6	10.1	10.0	9.2	8.2	7.7	7.6	7.8	8.5	8.8	9.0	8.8
MD	Baltimore	25	9.9	10.7	11.3	11.0	9.6	8.7	8.2	8.2	8.4	8.9	9.5	9.4	9.5
MA	Boston	18	14.2	14.2	14.0	13.4	12.2	11.3	10.8	10.8	11.4	12.2	13.1	13.8	12.6
MI	Detroit	41	11.6	11.5	11.5	11.1	9.9	9.1	8.3	8.1	8.9	9.5	11.3	11.3	10.2
	Sault Ste. Marie	34	10.1	10.0	10.5	10.8	10.2	8.9	8.3	8.1	9.0	9.6	10.2	10.0	9.6
MN	Duluth	26	12.1	11.9	12.1	13.4	12.3	10.8	9.9	9.8	10.7	11.5	12.2	11.4	11.5
	Minneapolis-St. Paul	37	10.4	10.6	11.3	12.4	11.4	10.5	9.3	9.1	9.9	10.5	11.0	10.3	10.6
MS	Jackson	12	9.1	9.0	9.6	9.1	7.3	6.4	6.1	5.8	6.7	6.5	7.9	8.7	7.7
MO	Kansas City	3	10.7	11.7	11.7	11.8	9.6	9.4	7.9	8.6	8.3	10.0	11.6	10.9	10.2
	St. Louis	26	10.3	10.8	11.8	11.4	9.4	8.6	7.6	7.4	7.9	8.5	9.9	10.2	9.5
MT	Great Falls	34	15.9	14.8	13.5	13.2	11.5	11.4	10.3	10.5	11.6	13.8	15.0	16.1	13.1
NE	Omaha	40	11.1	11.5	12.7	13.2	11.4	10.5	9.1	9.2	9.7	10.1	11.2	10.8	10.9
NV	Reno	33	6.0	6.1	7.6	8.0	7.6	7.2	6.6	6.2	5.4	5.3	5.4	5.1	6.4
NH	Concord	34	7.3	7.9	8.2	7.9	7.0	6.3	5.6	5.3	5.4	5.9	6.5	7.0	6.7
NJ	Atlantic City	17	11.8	12.2	12.5	12.3	10.8	9.7	9.1	8.7	9.3	9.7	11.1	11.3	10.7
NM	Albuquerque	36	8.0	8.8	10.0	10.9	10.5	10.0	9.1	8.1	8.5	8.3	7.8	7.7	9.0
NY	Albany	37	9.8	10.3	10.6	10.5	9.1	8.1	7.3	6.9	7.3	7.9	8.9	9.1	8.8
	Buffalo	36	14.5	14.1	13.8	13.0	11.8	11.3	10.6	10.1	10.6	11.4	13.0	13.4	12.3
	New York*	56	10.7	10.9	11.1	10.5	8.8	8.1	7.7	7.7	8.1	9.0	9.9	10.4	9.4
NC	Charlotte	26	8.0	8.5	9.0	9.1	7.6	6.9	6.6	6.5	6.8	7.1	7.3	7.4	7.6
	Raleigh	26	8.7	9.2	9.6	9.4	7.9	7.1	6.8	6.5	7.0	7.3	7.9	8.2	8.0
ND	Bismarck	36	10.1	10.1	11.3	12.6	12.2	11.0	9.6	9.9	10.4	10.2	10.4	9.6	10.6
OH	Cincinnati	43	8.3	8.4	9.0	8.4	6.7	6.4	5.2	5.1	5.4	6.1	7.7	7.9	7.1
	Cleveland	34	12.5	12.3	12.5	11.9	10.4	9.5	8.7	8.4	9.1	10.0	12.1	12.3	10.8
	Columbus	26	10.3	10.5	10.8	10.2	8.6	7.5	6.7	6.4	6.8	7.6	9.5	9.8	8.7
OK	Oklahoma City	27	13.3	13.7	15.1	15.1	13.3	12.7	11.3	10.8	11.5	12.2	12.7	12.8	12.9
OR	Portland	27	10.1	8.8	8.3	7.2	6.9	6.9	7.4	7.0	6.4	6.4	8.4	9.6	7.8
PA	Philadelphia	35	10.3	11.1	11.5	11.1	9.7	8.8	8.1	7.9	8.3	8.9	9.7	10.1	9.6
	Pittsburgh	23	10.7	11.0	11.1	10.8	9.3	8.2	7.5	7.3	7.7	8.4	10.1	10.5	9.4
RI	Providence	22	11.5	11.9	12.4	12.5	11.1	10.1	9.5	9.5	9.6	9.7	10.6	11.0	10.8
SC	Columbia	27	7.1	7.7	8.4	8.7	7.0	6.4	6.5	6.0	6.2	6.1	6.5	6.6	6.9
SD	Sioux Falls	27	11.0	11.2	12.6	13.6	12.1	10.7	9.7	9.8	10.3	10.8	11.6	10.7	11.2
TN	Memphis	27	10.6	10.6	11.3	10.8	9.0	8.1	7.6	7.1	7.6	7.8	9.4	10.1	9.2
	Nashville	34	9.2	9.4	10.0	9.7	7.6	7.0	6.4	6.1	6.4	6.5	8.4	8.8	7.9
TX	Dallas	22	11.5	12.3	13.3	13.1	11.4	11.0	9.7	9.3	9.7	9.9	10.9	11.2	11.1
	El Paso	33	9.2	10.0	11.9	12.0	11.0	10.1	8.9	8.4	8.4	8.2	8.6	8.6	9.6
	Houston	6	8.1	8.6	9.4	9.5	7.8	7.3	6.3	5.1	6.7	6.3	7.9	7.7	7.6
UT	Salt Lake City	46	7.7	8.2	9.2	9.5	9.3	9.3	9.4	9.5	9.0	8.5	7.8	7.5	8.7
VT	Burlington	32	9.7	9.4	9.3	9.3	8.8	8.2	7.8	7.4	8.0	8.6	9.5	9.7	8.8
VA	Norfolk	27	11.7	12.1	12.5	11.9	10.3	9.6	8.8	8.7	9.6	10.4	10.7	11.1	10.6
	Richmond	27	7.9	8.6	9.0	8.9	7.8	7.2	6.7	6.3	6.6	6.8	7.4	7.5	7.6
WA	Seattle-Tacoma	27	10.4	9.9	10.2	9.8	9.2	9.0	8.4	8.1	8.3	8.9	9.4	10.0	9.3
	Spokane	28	9.0	9.1	9.5	9.7	8.8	8.8	8.2	8.0	8.1	8.2	8.3	8.8	8.7
WV	Charleston	28	7.6	7.9	8.6	7.9	6.3	5.6	5.0	4.5	4.8	5.3	6.9	7.2	6.5
WI	Milwaukee	35	12.9	12.8	13.2	13.2	12.0	10.5	9.7	9.6	10.7	11.5	12.8	12.5	11.8
WY	Cheyenne	18	15.8	15.3	14.9	15.0	13.1	11.8	10.5	10.8	11.5	12.5	13.8	15.0	13.3
PR	San Juan	20	9.3	9.3	9.7	9.4	8.7	9.0	9.9	9.2	7.6	6.9	7.7	8.9	8.8

Source: U.S. National Oceanic and Atmospheric Administration, *Comparative Climatic Data.*
* City office data.

61

thereby indirectly affects the design of the system equipment which connects them. The direct effect is one of heat loss from building and collector due to convection and air leakage. A location with a relatively high average wind speed can be expected to suffer greater losses than one of relative calm. Average wind speeds for selected cities are provided in Table 3–10.

Wind must also be considered, since its velocity can damage or destroy the solar collection apparatus and its supporting structure unless strength equal to the expected loads is provided. The criteria for the design of structures and component parts of structures are presented in ANSI A58.1, Building Code Requirements for Minimum Design Loads in Buildings and Other Structures. The following procedure is outlined for developing design specifications.

1. Select the mean recurrence interval. This is generally 25, 50, or 100 years, with the latter interval providing the most severe wind speeds and most conservative design.

2. Select the wind speed appropriate for the general geographical location and any special local conditions.

3. Determine the effective velocity pressure, q_p, for the structure at the appropriate exposure, height, and selected wind speed.

TABLE 3–11 Collector Structural Design Rating Requirements
(pounds per square foot)

Exposure A: Basic wind speed (mph) for centers of large cities and very rough, hilly terrain

Height (ft)	60		100		130	
	$C_f = -1$	$C_f = 2$	$C_f = -1$	$C_f = 2$	$C_f = -1$	$C_f = 2$
30	−5*	10	−13	26	−21	42
200	−10	10	−29	58	−49	98
800	−20	40	−57	114	−96	192

Exposure B: Basic wind speed (mph) for suburban areas, towns, city outskirts, wooded areas and rolling terrain

Height (ft)	60		100		130	
	$C_f = -1$	$C_f = 2$	$C_f = -1$	$C_f = 2$	$C_f = -1$	$C_f = 2$
30	−8	16	−28	56	−40	80
200	−15	30	−42	84	−72	144
800	−24	48	−67	134	−113	226

Exposure C: Basic wind speed (mph) for flat open country, open flat coastal belts and grasslands

Height (ft)	60		100		130	
	$C_f = -1$	$C_f = 2$	$C_f = -1$	$C_f = 2$	$C_f = -1$	$C_f = 2$
30	−14	28	−38	76	−64	128
200	−20	40	−57	114	−96	192
800	−28	56	−76	152	−129	258

* Negative design ratings refer to suction-type loads.

4. Determine the maximum positive and negative pressure coefficients, C_f, appropriate for the tilt angle, critical wind direction, and type of supporting enclosure. The range of C_f is approximately -1 to 2.0 for common collector mounting designs. Negative coefficients refer to a suction-type load.

5. Calculate the $q_p C_f$ product to determine the structural rating required at the specific load per unit area.

6. The load per unit area multiplied by the gross collector area will result in the total design load to be used for mounting hardware and for the balance of the structural design.

Table 3–11 summarizes rating requirements for a variety of exposures, wind speeds, heights and pressure coefficients. The numbers in orange indicate those applications with a wind load which requires a collector rated for 50 pounds per square foot.

OTHER FACTORS

Atmospheric dust and moisture, air pollution and pollens, fog or mist in valleys and coves, can all affect the performance of solar energy systems. But their effects, like other environmental problems, such as bird or insect interference and stercoraceous factors, are so transitory in nature or so easily remedied that they need not be considered here.

SUMMARY AND APPROXIMATIONS

Because climate is the prime external influence in many solar energy applications, it is vitally important that the planning of solar systems include precise calculations of the specific effects of the climate at the site. Temperature, cloud cover, humidity, and wind speed will all affect performance and may dictate the use of design refinements which might otherwise be unneeded. The use of a selective absorber can change from optional to obligatory if high system efficiency must be achieved under mediocre climatological conditions. These refinements are more costly but they can also be the difference between long-term cost effectiveness and impotence.

It is possible, however, to utilize averaged climatological data and generalized cost figures to arrive at a total system expenditure for general areas of the United States. It is worth repeating that these approximations cannot substitute for precise definitions of on-site realities, since within these general geographical zones significant climatological variations occur. But it can be helpful to identify a reasonable first approximation of the required collection area, energy savings, and system costs.

The DOE has made relevant approximations in publication SE-101, "Solar Energy for Space Heating and Hot Water," from which the following maps, charts and tables are taken. As a first approximation the contiguous United States has been divided into 12 solar climate zones which are shown graphically in Figure 3–15. For each of these zones, the approximate collector and storage tank requirements have been computed. The dimensions necessary to provide

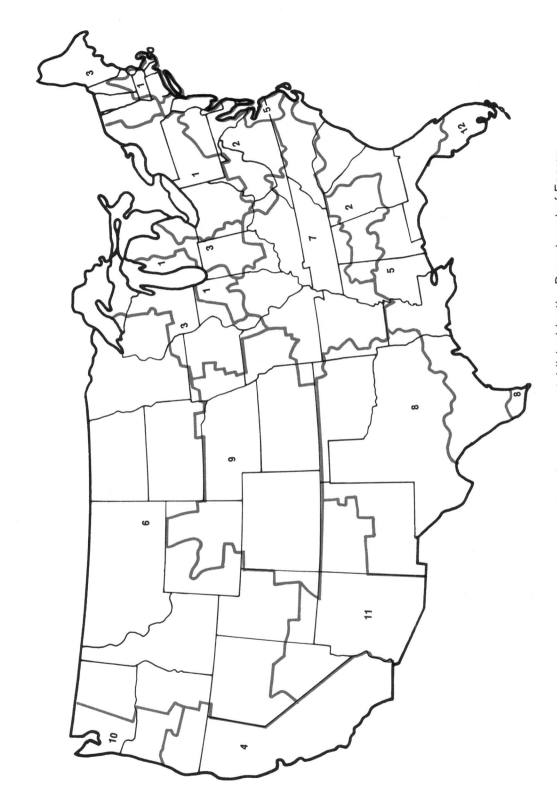

FIGURE 3-15 Solar climatic zones as established by the Department of Energy.

TABLE 3–12 Approximate Collector and Storage Tank Sizes Required to Provide the Heating and Hot Water Needs of a 1500-Square-Foot Home

Climatic Zone	Percent of Energy Supplied by Solar	Collector Area, Square Feet	Representative Collector Dimensions		Storage Tank Capacity, Gallons	Representative Cylindrical Storage Tank Dimensions	
			No. of 8-Ft.-High Rows	Length of Each Row, Ft.		Diameter, Inches	Length, Inches
1	71	800	3	33	1,500	48	200
2	72	500	2	31	750	42	138
3	66	800	3	33	1,500	48	200
4	73	300	1	37.5	500	48	78
5	75	200	1	25	280	42	60
6	70	750	3	31	1,500	48	200
7	70	500	2	31	750	42	138
8	71	200	1	25	280	42	60
9	72	600	2	37.5	1,000	48	132
10	53	500	2	31	750	42	138
11	85	200	1	25	280	42	60
12*	85	45	1	5.5	80	20	63

* Includes only hot water needs.

the indicated percentage of yearly heating and hot water energy demand for a typical 1500-square-foot home are shown in Table 3–12, those for a 10,000-square-foot building in Table 3–13. These figures are, as noted, only approximations and the dimensions of actual installation equipment can vary above or below the sizes shown. Storage tanks are especially variable, depending upon the defined objective of the system's operation. A good rule of thumb is that in a system using water for heat transfer and storage, about 1½ gallons of storage capacity should be available for each square foot of collector area. Heated air systems, which usually store energy in rocks, require about three times the volume storage capacity of a water system.

The performance of a generalized solar energy system in providing for heating and hot water requirements in a home of 1500 square feet is depicted by month for each climatological zone in Figures 3–16 through 3–18. Estimated total requirements are indicated in the right-hand bar for comparison with the left-hand bar which combines the incident solar energy on the indicated square footage of collectors tilted to an appropriate angle (approximately the local latitude angle) and the amount of energy which is actually collected. Note that for the months of November through March energy requirements are greater than solar energy collected, while during the remainder of the year collection usually exceeds demand. A greater proportion of winter heating needs could be met by installing more collection area but this adds significantly to costs. Instead, it is usually more economical to reduce heating requirements by installing more building insulation, storm windows, and other energy conserving measures. As is evident from the chart examples, solar energy supplies be-

TABLE 3-13 Approximate Collector and Storage Tank Sizes Required to Provide the Heating Needs of a 10,000-Square-Foot Building

Climatic Zone	Percent of Energy Supplied by Solar	Collector Area, Square Feet	Representative Collector Dimensions		Storage Tank Capacity, Gallons	Representative Cylindrical Storage Tank Dimensions	
			No. of 8-Ft.-High Rows	Length of Each Row, Ft.		Diameter, Inches	Length, Inches
1	74	5,330	7	95	10,000	96	326
2	71	3,330	5	83	5,000	72	300
3	68	5,330	7	95	10,000	96	326
4	73	2,000	4	62.5	4,000	72	246
5	75	1,210	3	50.5	2,000	54	215
6	72	5,000	7	80	7,500	84	322
7	71	3,330	5	83	5,000	72	300
8	74	1,330	3	55.5	2,000	54	215
9	75	4,000	6	83	6,000	72	354
10	60	3,330	5	83	5,000	72	300
11	77	1,000	3	41.5	1,500	54	163
12*							

* There is essentially no heating requirement in this zone.

tween 55 and 85 percent of the yearly heating and hot water requirements. Most solar systems are designed toward that range of supplying 50%–80% of energy demand, since this most often represents an acceptable compromise between costs and capacity.

The actual costs of solar energy systems, like the costs of any acquisition, are of prime importance. Too often, though, many purchases are made on the basis of initial cost with scant attention to the secondary but no less important costs of maintenance and operation. Unfortunately, those purchases where continuing costs can most easily become burdensome—homes, automobiles, air conditioners, and heating systems, among others—are most often those in which initial cost assumes an unwise ascendancy. Recognition of the importance of continuing costs is especially important with energy consuming items. A conventional heating or cooling system may require a lower initial investment than does a solar system but that advantage, extended over the continual use and long life of the equipment, is soon lost.

An illustration of the long-term superiority of solar systems is easily shown. Based upon the collector areas, heating and hot water requirements of Tables 3–12 and 3–13 for buildings of 1500 and 10,000 square feet, the energy savings for solar systems, which consume little external fuels, are given in Tables 3–14 and 3–15. The range of prices which a home or building owner can expect to pay for such systems—total costs which include design, materials and installation—are given in Table 3–16. The costs are not modest when viewed through the historic frame of energy expense, but after the economy of long-term operation costs, the inevitability of rising traditional fuel prices, and the expected decrease in solar costs are factored into the equation, solar systems

FIGURE 3–16 Performance of a generalized solar energy system: energy available, collected, required for a 1500-square-foot home.

67
Climatic Statistics and
Solar Insolation

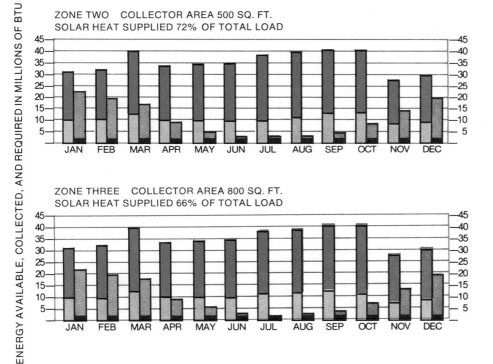

ZONE ONE   COLLECTOR AREA 800 SQ. FT.
SOLAR HEAT SUPPLIED 71% OF TOTAL LOAD

ZONE TWO   COLLECTOR AREA 500 SQ. FT.
SOLAR HEAT SUPPLIED 72% OF TOTAL LOAD

ZONE THREE   COLLECTOR AREA 800 SQ. FT.
SOLAR HEAT SUPPLIED 66% OF TOTAL LOAD

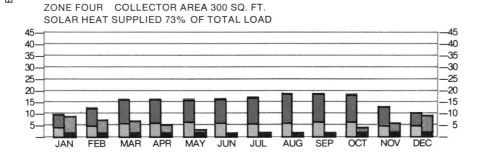

ZONE FOUR   COLLECTOR AREA 300 SQ. FT.
SOLAR HEAT SUPPLIED 73% OF TOTAL LOAD

ENERGY AVAILABLE, COLLECTED, AND REQUIRED IN MILLIONS OF BTU

Legend:
- SOLAR ENERGY AVAILABLE ON TILTED SURFACE
- ENERGY REQUIRED FOR SPACE HEATING
- SOLAR ENERGY COLLECTED
- ENERGY REQUIRED FOR HOT WATER

FIGURE 3–17 Performance of a generalized solar energy system.

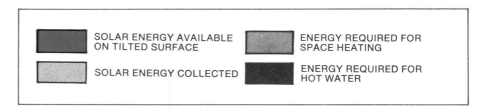

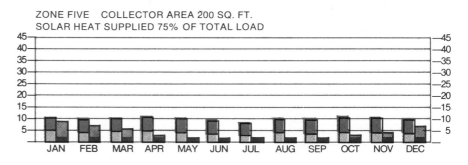

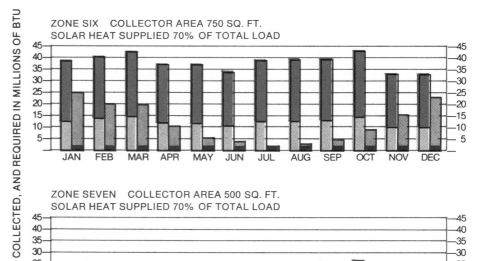

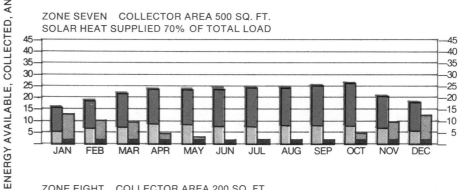

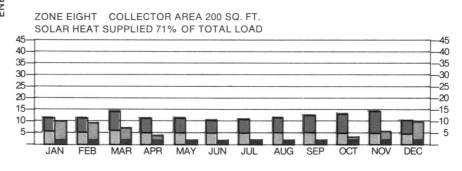

FIGURE 3–18 Performance of a generalized solar energy system.

SOLAR ENERGY AVAILABLE ON TILTED SURFACE

ENERGY REQUIRED FOR SPACE HEATING

SOLAR ENERGY COLLECTED

ENERGY REQUIRED FOR HOT WATER

ENERGY AVAILABLE, COLLECTED, AND REQUIRED IN MILLIONS OF BTU

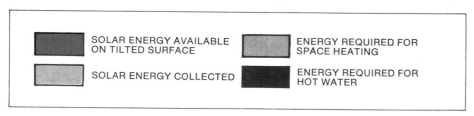

ZONE NINE   COLLECTOR AREA 600 SQ. FT.
SOLAR HEAT SUPPLIED 72% OF TOTAL LOAD

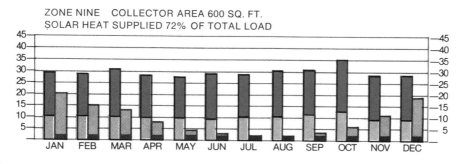

ZONE TEN   COLLECTOR AREA 500 SQ. FT.
SOLAR HEAT SUPPLIED 58% OF TOTAL LOAD

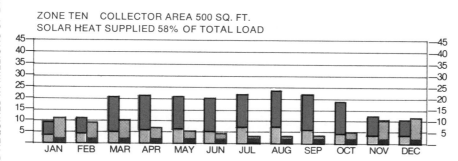

ZONE ELEVEN   COLLECTOR AREA 200 SQ. FT.
SOLAR HEAT SUPPLIED 85% OF TOTAL LOAD

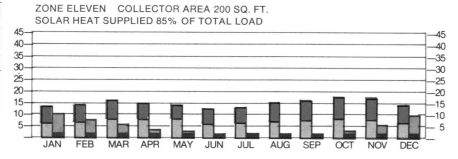

ZONE TWELVE   COLLECTOR AREA 450 SQ. FT.
SOLAR HEAT SUPPLIED 86% OF TOTAL LOAD

NOTE CHANGE IN SCALE

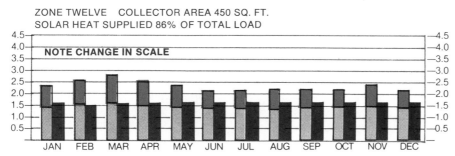

TABLE 3–14   Energy Supplied and Annual Dollar Savings
(1500-Square-Foot Building)

Climatic Zone	Energy Savings, Millions of BTUs	Comparison with Oil		Comparison with Electricity		
		Equivalent Gallons	Dollar Savings*	Equivalent Kilowatt Hours	Dollar Savings at Indicated Cost	Cost per kWh (Cents)
1	67.9	757	303	19,900	995	5
2	54.9	612	245	16,000	720	4.5
3	82.0	914	366	24,000	960	4.5
4	41.8	466	186	12,200	488	4
5	33.8	377	151	9,900	347	3.5
6	98.9	1,103	441	29,000	1,015	3.5
7	50.6	564	226	14,800	518	3.5
8	89.0	435	174	11,400	399	3.5
9	74.6	832	333	21,900	767	3.5
10	46.5	518	207	13,600	272	2
11	43.7	487	195	12,800	512	4
12	16.7	186	74	4,900	196	4

* 65% furnace efficiency at 40¢/gallon.

TABLE 3–15   Solar Energy Supplied and Annual Dollar Savings
(10,000 Square Foot Building)

Climatic Zone	Energy Savings, Millions of BTUs	Comparison with Oil		Comparison with Electricity		
		Equivalent Gallons	Dollar Savings†	Equivalent Kilowatt Hours	Dollar Savings at Indicated Cost	Cost per kWh (Cents)
1	372	4,147	1,659	109,000	5,450	5
2	268	2,988	1,195	78,500	3,530	4.5
3	472	5,262	2,015	138,300	6,220	4.5
4	186	2,074	830	54,500	2,180	4
5	128	1,427	571	37,500	1,310	3.5
6	587	6,544	2,618	172,000	6,020	3.5
7	253	2,821	1,128	74,100	2,594	3.5
8	174	1,940	776	51,000	1,784	3.5
9	418	4,660	1,864	122,500	4,290	3.5
10	242	2,698	1,079	70,900	1,420	2
11	165	1,839	736	48,300	1,930	4
12	*					

* There is essentially no heating requirement in this zone.
† 65% furnace efficiency at 40¢/gallon.

TABLE 3–16 Range of Solar Heating and Hot Water System Costs for 1,500 and 10,000-Square-Foot Buildings

Climatic Zone	Range of Solar System Costs for a 1,500-Square-Foot Home	Range of Solar System Costs for a 10,000-Square-Foot Building
1	$8,000–$24,000	$53,000–$159,000
2	5,000– 15,000	33,000– 90,000
3	8,000– 24,000	53,000– 159,000
4	3,000– 9,000	20,000– 60,000
5	2,000– 6,000	13,000– 30,000
6	7,500– 22,500	50,000– 150,000
7	5,000– 15,000	33,000– 90,000
8	2,000– 6,000	13,000– 30,000
9	6,000– 18,000	40,000– 120,000
10	5,000– 15,000	33,000– 99,000
11	2,000– 6,000	13,000– 39,000
12	450– 1,350	—

can be considered immediately competitive and, in the near future, superior. In addition, these figures represent the "arm chair" expenditures, i.e., the costs to be expected if the entire system installation is performed by a commercial contractor with no contribution by the home or building owner. There are any number of areas where substantial cost reductions may be effected through do-it-yourself efforts. The addition of more building insulation to reduce demand (and therefore equipment requirements) is perhaps the easiest method of reducing costs. Other contributions, depending upon the owner's skills and inclinations, such as the construction of collector plate support structures, the plumbing of pipe connections, and the installation of in-home delivery equipment, can produce larger economies.

Solar energy utilization, no less than the business of accomplishing a day's work, is a matter of making intelligent choices. The diversity of potential applications, each with differing demands and each operating under unique climatological conditions, can be met with an equal diversity of equipment, operational options to counter the most difficult problems, and a wide range of costs. Whether to use more costly materials to reduce collection area or to improve performance at the expense of greater weight are general questions addressed to specific problems. Making those decisions intelligently is not difficult once expertise is applied to the problem. The methods of precise delineation of system needs, exact determinations of environmental and operational parameters, and the implementation necessary for an economical and efficient system—in short, application expertise—are the subjects of the remainder of this handbook.

CHAPTER 4
COLLECTION TECHNIQUES

A summary list of solar energy collection techniques contains a diversity of entries, some of which are not a little surprising. One method has been in use for thousands of years while another, though perhaps not always recognized as a collection process, is practiced in construction planning. Each method has its own special advantages and disadvantages which must be considered before choosing the method most appropriate for a particular application.

Collection by *architectural rendering* consists of recognizing the effects of the sun in the architecture of a building so that maximum advantage can be incorporated into design features which might otherwise be determined arbitrarily or purely on aesthetic grounds. An appreciation of the possibilities has been a consideration in some construction planning for many years, often as a subsidiary concern, but the pressures of the contemporary fuel markets have forced a greater awareness. The orientation of buildings, the design of eaves on the glazed walls facing south, the choice of deciduous trees on the south and evergreens on the north, roof angles, interior coloration, and accommodation for thermal insulation are a few examples of architectural renderings which collect solar energy in a purposeful way. Each has value in reducing a building's energy demand when a conventional system is used (it is obviously self-defeating to cool a building while simultaneously allowing the sun to heat it through large areas of window glass) but each has greater importance in a solar system application.

Sole reliance on architectural rendering for energy supply is possible only in applications requiring very modest temperature changes, since it cannot boil water or generate electricity, but judicious architectural choices combined with features specific to solar systems, such as planning for equipment installation space, should form the first step of engineering for a solar energy system application. Architectural rendering is a *passive* technique, to be sure, and one which is still bound by aesthetic and cost considerations, but it can make a long-term contribution more than equal to the effort involved.

The *flat plate solar collector* is the most economical *active* method of collecting solar energy. It consists of an assembly of transparent covers over a collector plate which is backed with thermal insulation. Typically, a flat plate collector is used to heat water or air which is then utilized in hot water space heating and other services, or stored for later use. Because of the importance of the flat plate collector in solar energy applications, it is considered in greater detail later in this chapter.

Concentrating collectors deflect sunlight from a large area into a smaller region where the concentration of light can be used to yield temperatures higher than those obtainable from flat plates. Focusing mirrors or lenses are used to effect the concentration but they must be aligned more precisely toward the sun for greatest efficiency. This often requires a tracking system—electrical or mechanical techniques which, though more costly, yield greater energy collection than a stationary plate—to move the focusing mechanism in concert with the sun's motion.

Some concentration techniques can be used without tracking systems (the Winston design is one example), but there are practical and theoretical limits to the degree of concentration and, therefore, the efficiency that can be achieved without tracking.

Photovoltaic collectors convert sunlight directly into electricity through the action of the sun on a semiconductor junction to produce an electrical voltage which supplies current through conducting wires. These collectors, often called solar cells, can be incorporated into a flat plate structure or used as a conversion element in concentrating collectors to simultaneously produce both thermal and electrical energy for immediate use or for storage in thermal tanks and batteries.

In another application, a photovoltaic cell is used to protect bridges and other outdoor structures from electrochemical corrosion resulting from the contact potential which is produced between the metallic structure and the moist earth. The voltage provided by a photovoltaic solar cell mounted on the structure is used to counter the contact potential. As costs become more affordable, solar cells can be used to charge batteries which are used only intermittently: in boats, golf carts or lawn care equipment which may stand idle for long periods.

In summary, photovoltaic collectors provide the most versatile form of energy—electricity—but at present are cost effective in a very limited range of applications, such as powering space vehicles and in remote locations.

Bioconversion is a general category of solar energy utilization techniques wherein the sun powers the vegetative photosynthesis which converts CO_2 and water to combustible carbohydrates. Wood fuels supply energy stored by the living tree, an ancient source of energy but one which is founded on the sun. The method is inefficient since the energy yield of a combustion process is much less than the energy input to the plant, but the process has been improved.

Several fuel alcohols (methanol, ethanol) and gases (methane) can be distilled by synthetic or natural processes to be used as fuels, a possibility in underdeveloped nations where the availability of land otherwise unsuitable for plant food production and cheap labor resources can contribute to cost effectiveness. It is probably not an economical solar energy technique for industrialized nations.

Power generation through *natural geographic features* by using windmills, ocean thermal gradients, tides, and waterfalls are all solar sources, since the sun is the basis of provision for each of these opportunities. Though they are not direct utilizations of solar energy, they have the advantage of being as effective at night as they are during the day; but they depend upon natural formations for power generation at the site with the necessity for a distribution mechanism for supply elsewhere. Southern Florida has no waterfalls, Kansas has no tides, Los Angeles experiences long periods with little wind: all are obvious and severe limitations, like those attendant to geothermal power.

Extraterrestrial production of solar energy refers to the use of artificial earth satellites that would collect solar energy in space, convert it to microwave radio beams, and transmit it down to the earth. This technique has never been demonstrated but has been proposed by the Arthur D. Little Company, which holds a patent on the process.

Other extraterrestrial techniques could function by orbiting large automated process plants in highly elliptical orbits around the sun similar to the path of the minor planet Icarus (named for the legendary wingclad aviator whose wax-constructed wings melted when he flew too close to the sun). Such a process plant could convert low-grade meteoric ores into highly concentrated, scarce metals during its close path to the sun for delivery upon return to the vicinity of the earth. These concepts may seem to have the aroma of fantasy clinging to them, but perhaps no more so than the concept of television if introduced to those living a century ago.

FLAT PLATE COLLECTION

The primary active technique is the flat plate collector, an assembly of transparent covers, absorber plate, insulation, and a supporting frame, as illustrated in the sectional drawing in Figure 4–1. The covers serve three functions: to prevent convection losses from the collector to the air while transmitting the sun's rays, to reduce thermal radiation losses from the collector, and to protect the collector's absorber plate against environmental hazards. The specifics of a given site and its appropriate solar energy system will determine the precise cover configuration; some collectors will operate most efficiently with one, two or three covers, while others may need none. The absorber is a coated plate upon which the sun's energy is converted to heat. The choice of coating material is, again, a function of the demand which the system must serve and the on-site conditions under which it must operate, but all collectors conduct the heat generated on the surface to a site for transfer of the heat to either fluid or

FIGURE 4–1   Flat plate solar collector.

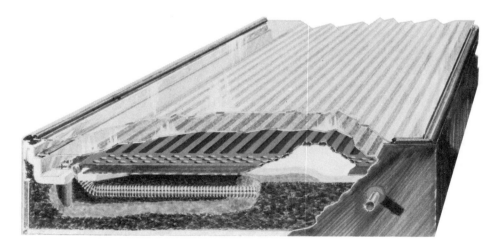

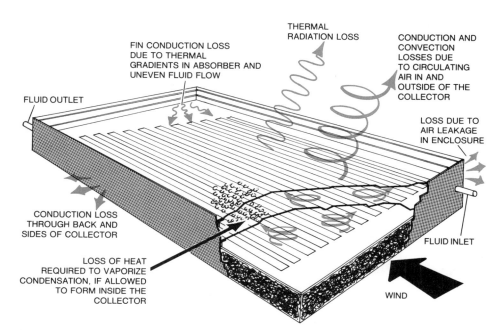

THERMAL
RADIATION LOSS

FIN CONDUCTION LOSS
DUE TO THERMAL
GRADIENTS IN ABSORBER AND
UNEVEN FLUID FLOW

CONDUCTION AND
CONVECTION
LOSSES DUE
TO CIRCULATING
AIR IN AND
OUTSIDE OF THE
COLLECTOR

FLUID OUTLET

LOSS DUE TO
AIR LEAKAGE
IN ENCLOSURE

CONDUCTION LOSS
THROUGH BACK AND
SIDES OF COLLECTOR

FLUID INLET

LOSS OF HEAT
REQUIRED TO VAPORIZE
CONDENSATION, IF ALLOWED
TO FORM INSIDE THE
COLLECTOR

WIND

FIGURE 4–2 Factors that contribute to losses in a flat plate collector.

air. Since heat absorption and transfer is the *raison d'etre* of the collector assembly, heat losses must be kept to a minimum. Most commercially produced collectors provide insulation behind the collector plate toward this end, but a well-designed assembly will use additional insulation between the edges of the plate and its frame to further reduce losses there. The frame performs the obvious function of holding the assembly together and, depending upon the construction, of providing support to the collector plate. In some designs a pan in which the insulation is mounted is attached to, or is a part of, the frame, while other collectors will use two frames: an inner frame to support the absorber plate and an outer frame to contain insulation, hold the covers, and provide further support. The objective of the entire assembly is to achieve energy gathering with a minimum of the potential loss factors illustrated in Figure 4–2.

COLLECTOR PLATE CONSTRUCTION

The principal methods used in the construction of collector plates for liquid heating are shown in Figure 4–3. Integral construction, shown in drawing (*a*), is usually the best choice for most solar systems. It provides the most efficient heat transfer, flexibility in fluid flow path design capability, can be mass produced, and uses the minimum weight of material per unit of total weight. Solder bond construction, Figure 4–3*b,* is the most easily constructed and can be made from readily available copper which will conduct heat and is easily soldered, points which have not been lost on the individual hobbyist, but it has poor thermal contact between plate and conduit, as does the wired construction (Fig.

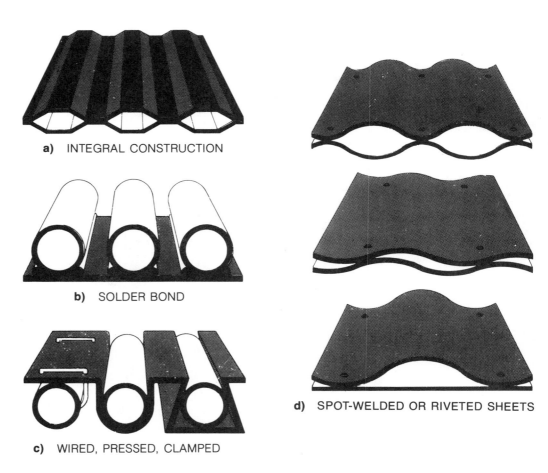

a) INTEGRAL CONSTRUCTION

b) SOLDER BOND

c) WIRED, PRESSED, CLAMPED

d) SPOT-WELDED OR RIVETED SHEETS

FIGURE 4–3 Collector plate construction.

4–3c). Pressed and clamped constructions increase thermal contact but they use much more metal, suffer corrosion under moist conditions when dissimilar metals are used, and are usually limited to linear conduit configurations. The spot-welded construction in Figure 4–3d has good thermal contact characteristics and is economical since it uses a minimum of material, but all edges must be sealed in a separate operation and its longevity is limited by the susceptibility of spot-welds to fail as tension and temperature cycling take their effects.

Collector plates, the actual site of heat absorption, offer an equally wide choice of construction materials. Rubber and plastics impregnated with carbon black will perform satisfactorily in simple installations such as swimming pool heaters, but those systems under greater demand require the greater efficiency obtained by construction with metals of high thermal conductance. Copper, aluminum, and steel qualify and are the most common collector plate materials but none of the three is without disadvantage. Copper has the highest thermal conductance but is quite expensive. Steel is usually more economical but is heavy and may rust. Aluminum is low in weight but can corrode when in contact with other metals. Any problem inherent in a particular material can be overcome or minimized so the ultimate choice of collector material must be made in response to the special conditions of each application.

TRICKLE COLLECTORS

Flat plate collectors may be designed so that the water dribbles down the face of the collector rather than flowing through enclosed tubes—a much more simple construction, but one with limited use. Corrugated metal, as shown in Figure 4–3*d,* is often used in this construction since the corrugations provide ready-made flow paths. Or the fluid may be directed, as pictured in Figure 4–4, in a slightly more complex manner. Both flow configurations are referred to as "open-channel flow," but there are serious limitations to the effectiveness of all trickle collectors. The action of water can damage the absorbing surface and, conversely, the surface can contaminate the water with a consequent reduction in the available options for absorptive coatings. Free moisture between the plate and transparent covers is unavoidable with open-channel flow, since the fluid alternately evaporates from the hot absorber surface and condenses on the underside of the cover; this results in increased convection losses and partial blockage of sunlight transmitted through the cover. Also, the flow in trickle collectors must be directed downward, requiring greater pumping effort rather than taking advantage of the siphoning effect and the natural tendency of a heated fluid to rise in a closed loop. Despite their limitations, these variously named "trickle plates" or "dribble collectors" are low in initial cost and, when the additional area and mechanical support for them is available, can be used to advantage.

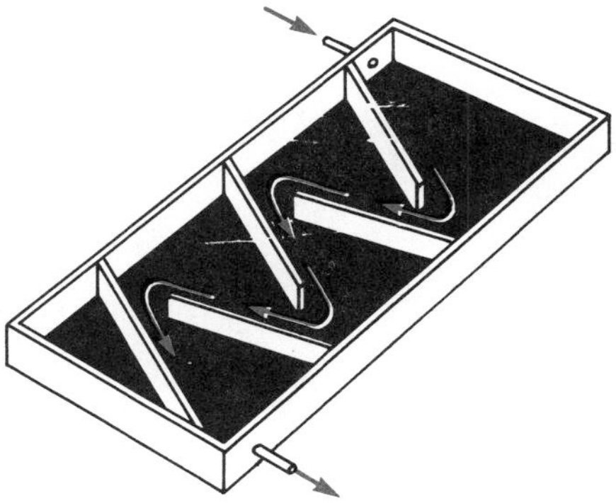

FIGURE 4–4 A low-cost but inefficient trickle collector.

FLOW CONFIGURATIONS IN CLOSED CONDUITS

The flow through a collector plate can be either in series or in parallel configurations, as compared in Figure 4–5. Differing characteristic temperature and pressure gradients provide several points of comparison which are worthy of note.

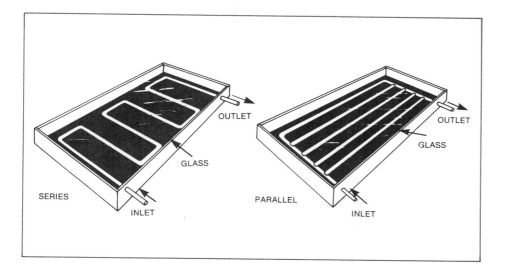

FIGURE 4–5 Series and parallel flow configurations.

TABLE 4–1 Tube Spacing for 95% Efficiency

Collector Plate Thickness (mm)	Collector Plate Material Tube Spacing (cm)		
	Copper (w)	Aluminum (w)	Steel (w)
0.1	9.0	7.5	4.0
0.2	11.5	9.5	6.0
0.3	12.5	10.5	6.5
0.5	14.5	13.0	7.5
1.0	16.0	15.0	9.5
1.5		16.5	10.5
2.0		18.0	11.5
3.0			13.0

If all the tubes are soldered to the same plate, the series configuration yields a non-uniform heating pattern since the fluid is heated as it progresses along the first path, then is cooled as it passes back toward the inlet region of the plate. In the parallel configuration, by contrast, fluid temperature rises continuously as it proceeds along parallel paths. In addition, the parallel configuration is more advantageous because the flow rates in the paths are self-adjusting by the thermosiphon effect, i.e., the tendency of a fluid to rise as heat reduces its density. This upward motive effect is of sufficient magnitude in a thermosiphon system (detailed in a later section) that no additional pumping is required to move the fluid. An upward directed flow is desirable in either configuration but, with all else equal, the higher average collector plate temperature gradient of the series configuration make it less efficient.

The series configuration also requires a longer flow path than does the parallel, with a number of related consequences. Given the same flow rate at the inlet, and equivalent tubing diameters in both configurations, the series path has a greater fluid velocity, with more resistance to flow, and therefore a greater pressure loss across the collector. Minimizing this problem is best accomplished in the planning stage of the solar system by calculating and comparing the values for pressure loss for each of the collectors under consideration for the application. The pressure difference between the inlet and outlet of a tube length, L, and inside diameter, d, is expressed in terms of the equivalent head loss, h, in units of length. The head loss, h, is identical to the pressure produced by a column of water of height, h.

With an open outlet the head loss, h, is approximated by:

$$h = \frac{32VL\mu}{g\rho d^2} + (h_0 - h_i) \tag{4-1}$$

where: h is the head loss across the collector (ft)

h_0 is the height of the outlet (ft)

h_i is the height of the inlet (ft)

V is the velocity of fluid in the tube (ft/sec)

L is the length of the tube (ft)

μ is the dynamic viscosity of the fluid (lb-sec^2/ft)

d is the diameter of the tube (ft)

ρ is the mass density of the fluid (lb/ft^3)

g is the acceleration of gravity $= 32.2$ ft/sec^2.

The units may be either in the metric or English system, but must be consistent. In a closed loop system, the $(h_0 - h_i)$ term is zero and drops out of the equation.

Tube spacing in a flat plate collector depends upon the tubing diameter, thermal conductivity and thickness of the absorber plate, and the heat transfer coefficients between both fluid and plate and plate and surroundings. An approximate guide to the tube spacing sufficient to provide an efficiency of 95% in the transfer of heat from the absorber plate to the fluid is shown in Table 4–1.

A useful expression for the fin efficiency of an absorber plate is derived by Duffie and Beckman:

$$F = \frac{\tanh\left[\sqrt{\dfrac{U_L}{k\delta}}\,\dfrac{(w - u)}{2} \right]}{\sqrt{\dfrac{U_L}{k\delta}}\,\dfrac{(w - u)}{2}} \tag{4-2}$$

where: F is the fin efficiency, expressed as a number between 0 and 1

  U_L is the overall loss coefficient of the absorber plate from Equation 4–12 in BTU/hr-ft^2-°F

  k is the thermal conductivity of the absorber plate in BTU/hr-ft^2-°F-in

  δ is the thickness of the absorber plate in inches

  w is the distance between centers of fluid conduits in inches

  u is the diameter of the fluid conduit in inches.

Air-heating flat plate collectors offer some theoretical advantages over collectors which use a liquid as the heat transfer medium. Among those advantages are potential cost savings since less plumbing is required, the complications of leakage of a liquid transfer medium (though not of thermal losses) are ameliorated, the possibility of plumbing corrosion is reduced, and the collector itself may be of a more simple construction. Further, the demands of choosing and constructing the heart of the collector—its absorber surface—are less stringent, since there is no need for integral conduits or the necessity to conduct heat from one area of the absorber to another. Because any blackened absorber surface will transfer heat to air blown over it, though with varying degrees of effectiveness, there are any number of possibilities for use an an absorber. Several of those possibilities, black gauze, glass plates, and metal plates with fin backing, as well as various construction configurations for air-heating collectors, are illustrated in Figure 4–6.

Air flow characteristics through the collector are of importance. Normally, air will flow in a laminar fashion, i.e., a layering of the air moving through the collector with the layer nearest the absorber plate absorbing most of the heat

FIGURE 4–6   Flat plate solar collectors for air heating.

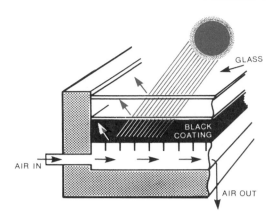

a)   FIN BACKING ON COLLECTOR PLATE

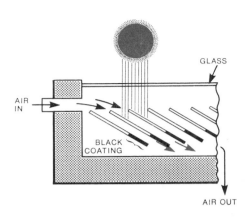

b)   LOUVERED COLLECTOR PLATES

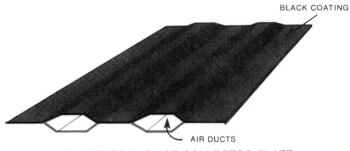

c)   INTEGRAL DUCT COLLECTOR PLATE

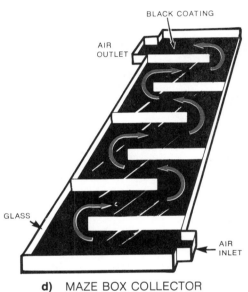

d)   MAZE BOX COLLECTOR

e)   POROUS BLACK GAUZE COLLECTOR

and relatively less heat transfer and cooler temperatures in the overlying layers. A better choice is to cause turbulent air flow where these layers are forced to mix as they proceed through the collector. Corrugations in the plate, fins, louvers, or the partitions of a maze box collector all serve to achieve turbulence. At the same time, however, over-restriction of the air flow in an effort to achieve turbulence can have deleterious effects, notably a greater pressure drop across the collector. In many cases, this situation will necessitate the use of a larger fan which, in turn, may require energy consumption greater than that saved by the solar energy system. Another general point worth notice on the matter of air flow is that better designs of air-heating collectors direct the air flow away from the transparent covers after heating by the absorber plate.

Damage from leakage of liquid transfer medium from system plumbing is not a concern with air-heating collectors, but leakage remains a serious concern. Because the air flowing through the system is under pressure, any fault in the system's integrity, however small, will reduce collector efficiency. Loss of solar-heated air through cracks in the collector or ductwork and, conversely, infiltration of cooler air anywhere in the system, can be crucial. In addition, though not strictly a leakage problem, the necessary turbulence of the air flow in the collector causes a relatively larger thermal loss through convection than is the case with a similar liquid transfer unit.

Air-heating collectors have a number of disadvantages as well. The most obvious is that they cannot be effectively used to provide cooling or hot water service. For reasons noted in a previous chapter, the inability to efficiently provide domestic hot water is a serious fault, since hot water service will probably prove to be the object in the initial stage of growth of solar energy applications. Combining an air-heating collector to provide space heating with photovoltaic cells to generate electricity for cooling and/or hot water service is one possibility but the availability of these cells at acceptable cost levels remains only a future possibility.

Another problem in using air rather than liquid for heat transfer is the significantly lower capacity of air to contain heat. The specific heat of air is 0.24, while the specific heat of water is 1.0, more than four times greater. Further, the lower density of air (approximately 0.075 pounds per cubic foot) as compared to water (more than 62 pounds per cubic foot) carries with it other consequences. Much greater quantities of air in both volume and weight must be moved through the collector to transfer an equivalent amount of thermal energy. Both of these factors, in turn, necessitate the use of increased spaces through which to move the air, not only in the collector but in the service ductwork, and greater surface area in ductwork can mean greater thermal losses from that surface. Larger ductwork space within the collector will decrease the pressure drop across the collector but, at the same time, reduce heat transfer between plate and air. And, of course, larger ductwork requires more expense in materials. As is the case so often in solar system design, and as will be shown throughout succeeding chapters, final system design is usually a product of trade-offs in many areas. It is reasonable to project, however, that unless a second energy-generating system (such as a photovoltaic cell) is combined with it, the air-heating collector will find utility only in applications such as crop drying, home heating, or preheat for plant air make-up where space heating is the lone objective.

TRANSPARENT COVERS

The cover of a solar collector protects the absorber plate from environmental hazards, but its primary function is to reduce the heat loss upward by thermal convection. Collectors can be protected by several successive covers but, while each addition reduces convection loss to a greater degree, each additional cover will also reduce the amount of solar energy admitted because of reflectance from, and absorption by, the added layers. The problem can be attacked by considering solar energy admission as most important, but the magnitude of convection loss is usually more critical. The rate of this heat loss increases with wind velocity, the difference between the collector plate and ambient temperatures, and the tilt angle, Σ, of the collector. But, assuming a given standard value for each of these variances, it is possible to represent the effect on collector efficiency by additional covers alone.

Figure 4–7 shows a typical curve representing the efficiency of a flat plate collector as a function of the average fluid temperature for one, two, and three transparent covers. Note that for a low fluid temperature, greater efficiency is obtained with one cover than with several. At higher fluid temperature re-

FIGURE 4-7 Effect of glazing covers on ASHRAE 93-77 data for black chrome and flat black paint at 40° tilt angle, 50°F ambient temperature, and $\Theta = 0°$.

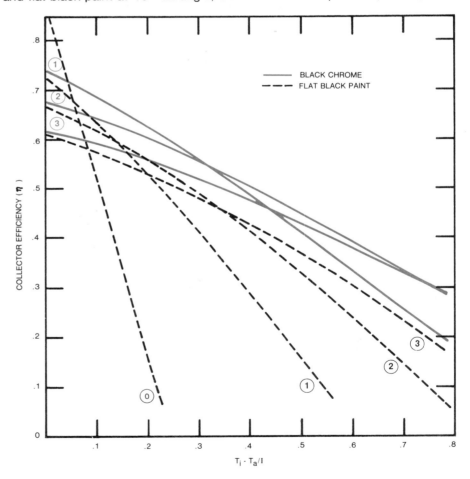

quirements, the achievement of greater efficiency requires an increasing number of covers. The heat loss upward through the transparent covers of a flat plate collector can be approximated from a formula by S. A. Klein:

$$J_{up} = \left[\frac{\dfrac{T_p - T_a}{N}}{\dfrac{C}{T_p}\left(\dfrac{T_p - T_a}{N + f}\right)^{.33} + \dfrac{1}{h_w}} \right] +$$

$$\left[\frac{\sigma(T_p^4 - T_a^4)}{\left(\dfrac{1}{E_p + .05N(1 - E_p)}\right) + \left(\dfrac{2N + f - 1}{E_p}\right) - N} \right] \quad \textbf{(4–3)}$$

where: J_{up} = frontal heat loss (BTU/hr-ft^2)

T_p = plate surface temperature (°R) (°R = °Rankine = °F + 460)

$\cong (T_o + T_i)/2$ for efficient plates

T_o = outlet temperature (°R)

T_i = inlet temperature (°R)

T_a = outdoor ambient temperature (°R)

N = number of cover plates

E_p = emissivity of the absorber plate

E_p = emissivity of the cover plate (0.88 for glass)

σ = 1.7132 × 10^{-9} BTU/ft^2-hr (°R)^4—the Stefan-Boltzman constant in English units. See Equation A–2

h_w = 1 + 0.3V BTU/hr-ft^2 °R where V is the wind volocity in mph

f = (1 − 0.227 h_w + 0.0161 h_w^2) (1 + 0.091N)

C = (95.556) (1.0 − 0.00883s + 0.0001298s^2)*

s = tilt angle from horizontal (degrees).

The standard deviation of the difference in values of J_{up}/A $(T_p - T_a)$ for 972 observations was 0.14 W/M^2 °K (0.0247 BTU/ft^2 °R).

The limits of the testing conditions were:

$116°F \leqslant T_p \leqslant 296°F$

$8°F \leqslant T_a \leqslant 98°F$

$0.1 \leqslant E_p \leqslant 0.95$

$0 \leqslant V \leqslant 22\,\text{mph}$

$1 \leqslant N \leqslant 3$

$0 \leqslant s \leqslant 90°$

With the calculated value of convection heat loss for each of the alternative cover configurations, the optimum choice of one, two, or more covers for a specific application may be made. The secondary concern associated with increasing the number of covers—an increasing reduction in solar energy admitted to the collector—may be addressed if necessary. The usual practice, how-

* Empirical testing by AMETEK yielded C = (115.3) (1.0 − 0.00409s). Parameters were 40° to 62° tests.

ever, is for collector manufacturers to supply an efficiency curve for their product and its cover configuration. The specific characteristics of the cover material, and the effect of the covers are accommodated in the curve which is sufficient data for an evaluation of equipment for a particular application. At the same time, it is well to be aware of the relationships from which the transmittance, τ, absorptance, α, and reflectance, ρ, of various cover materials may be calculated, as well as the means of quantifying the effect of internal reflections between covers in a multiple cover configuration. The equations and a discussion of their use are contained in Appendix B. As a general rule, covers should be constructed of materials which possess a high transmittance, τ, for solar radiation, low reflectance, ρ, low absorptance, α, and, of course, should be able to withstand environmental hazards, such as the impact of hailstones and high winds.

The incoming solar energy that is absorbed on the collector plate of a flat plate collector is approximately given by:

$$J_{in} = \alpha_e \tau_e I_T = (\alpha\tau)_e I_T \tag{4-4}$$

where: J_{in} is the power absorbed by the absorber plate in BTU/hr-ft^2

α_e is the effective absorptance of the absorber plate at solar wavelengths

τ_e is the effective transmittance of the transparent covers at solar wavelengths, given by Equations B-3, B-5 or B-8, whichever is appropriate

I_T is the total solar insolation given by Equation 3-1.

The quantity $\tau_e \alpha_e$ is referred to as the effective transmittance absorptance product, denoted $(\tau\alpha)_e$.

Further refinement of the analysis of incoming solar energy requires that multiple reflections between a diffusely reflecting absorber plate and its covers be considered. This requires knowledge of the dependance of the reflectance of the absorber plate on incident angle and the degree to which it is specular or diffused, points which are also discussed in Appendix B. When absorber reflectance is minimized, and it is desirable to do so, the effects of such multiple reflections are very small.

Another refinement consists of dividing the incoming total solar insolation into its direct beam, diffuse, and reflected components. In principle, Equation 4-4 should only apply to the direct beam radiation, but the distribution of diffuse and reflected radiation with various incident angles is complex; unless some assumptions are made it is difficult to determine its effective incident angle. One approach often used is to assume that diffuse radiation is isotropically distributed (equally strong from all directions) and use an average angle of 60° for its incident angle. In fact, diffuse radiation is not isotropic. Its intensity tends to increase toward the zenith (straight up) for that which emanates from clouds, while the intensity of the blue sky diffuse radiation (Rayleigh scattering) is greatest in the direction of the sun. Moreover, diffuse and reflected radiation is polarized in varying degrees and has varying spectral characteristics.

It is often satisfactory to use values of α and τ that are averaged over all values of incident angles and averaged over all solar wavelengths in Equation 4-4. This is actually required for the diffuse and reflected components of I_T.

The effects of polarization at large incident angles are usually ignored, as are the effects of variations in the index of refraction of air with temperature and humidity.

THE INCIDENT ANGLE MODIFIER

Since the effective transmittance absorptance product $(\tau\alpha)_e$ varies with the incident angle, θ, and depends in a complicated manner on the properties of, and number of, transparent covers, a short cut has been devised to describe the variation of $(\tau\alpha)_e$ with incident angle.

The short cut involves defining an incident angle modifier as the ratio of $(\tau\alpha)_e$ at incident angle, θ, to its value at normal incidence, $\theta = 0$. The incident angle modifier is:

$$K_{\tau\alpha} = \frac{(\tau\alpha)_e}{(\tau\alpha)_{e,n}} \qquad (4-5)$$

It has been shown that the incident angle modifier can be approximated by the expression:

$$K_{\tau\alpha} = 1 - b_0 \left[\frac{1}{\cos\theta} - 1 \right] \qquad (4-6)$$

for most flat plate collector geometries and for values of θ that are within the usual operating region of practical interest (i.e., not too close to 90°).

The determination of this relationship when $\Delta T/I = 0$ is one of the requirements of the ASHRAE 93-77 test procedure. This procedure simultaneously measures the reductions due to reflections off the surface of both the cover glazing and the absorber panel, and the additional effect of the edge shadow modifier (which is discussed separately below). The brief discussion of the edge shadow modifier is included here, but no separate testing or calculation of its effect is necessary when the ASHRAE 93-77 procedure is used.

THE EDGE SHADOW MODIFIER

Since the absorber plate of a flat plate collector lies below the level of the outermost transparent cover, the edges of the collector which support the covers will cast shadows onto the absorber plate. This effect can be significant at high incidence angles, especially if the collector is of a deep profile.

The width of the shadows cast onto the absorber surface can be determined from Equations 2–32 and 2–33 for the usual collector orientations on tilted surfaces. The value to be used for the "height of the fence, Γ," is the height of the collector edge above the absorber plate.

The edge shadow modifier for absorber plates of the same dimensions as the aperture is given by:

$$K_s = 1 - \frac{L_c w_1 + W_c w_2 - w_1 w_2}{A_c} \qquad (4-7)$$

where: L_c is the length of the collector aperture

W_c is the width of the collector aperture

$A_c = L_c W_c$ is the area of the collector aperture

w_1 is the shadow width from Equation 2–32

w_2 is the shadow width from Equation 2–33.

The edge shadow modifier is also multiplied by the direct beam solar insolation in order to account for the shaded area on the absorber plate at incident angles greater than zero.

There is some benefit from the solar insolation that strikes the inner wall of the edges opposite to those that shade the absorber plate. For this reason, though the benefit is small compared to the shadow loss, it is best for those visible inner walls to be reflecting or white in color so that the greater portion of this solar energy will be reflected onto the absorber plate. The loss due to shadows is greater than the simple geometrical loss of solar insolation, since the forced circulation of fluid in the shaded region of the plate continues to deliver heat to this cooler portion, resulting in a negative contribution to net energy delivered.

CONVECTION SUPPRESSION WITH HONEYCOMBS

Honeycomb devices consisting of planar arrays of closely packed hexagonal or square cells may be used to suppress convection currents which contribute to the upward losses in flat plate collectors. K. G. T. Hollands reported in 1965 on the effectiveness of honeycomb structures in this application.

FIGURE 4–8 The critical temperature difference below which convection is suppressed by a transparent honeycomb structure. (After K. G. T. Hollands)

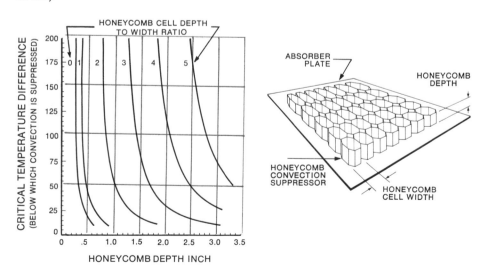

Honeycomb structures will suppress convection currents up to a critical temperature difference that depends upon the honeycomb depth and the depth-to-width ratio. Figure 4–8 shows this relationship as determined by Hollands for a square honeycomb array.

The heat transfer coefficients across a honeycomb panel include terms for convection, conduction, and radiation. If the honeycomb walls are very thin and highly transparent to solar radiation, conduction and radiation losses are minimized. A reduction in convection loss by nearly 80% is theoretically possible with an idealized honeycomb structure but this is not achieved because of the added conduction and radiation losses contributed by the honeycomb structure itself. Typically, the reduction in net upward loss is from 30% to 70%, depending upon the numerous factors which must be considered and the operating conditions in which the improvement is calculated or measured.

Honeycomb structures are currently being investigated for their effect on the performance of solar collectors. A firm conclusion upon their efficacy must await the resolution of questions which remain, not the least of which are those related to the high-temperature stability of the low-cost plastic materials which are the most easily fabricated into honeycomb structures.

VACUUM CONVECTION SUPPRESSION

Convection loss can be very effectively suppressed by removing the air altogether from the space between the absorber plate and the transparent covers. Evacuated collectors have been conceived and built using posts to support the cover against collapse due to the pressure of the outside atmosphere. In addition, long, evacuated glass tubes have been demonstrated to operate efficiently as transparent covers when used to enclose black absorber surfaces. Arrays of evacuated glass tubes offer advantages of good performance but high costs and the difficulties of the numerous vacuum seals which are required in large planar areas of collectors leave some room for further improvement.

Evacuated tube collectors, because of the circular cross section of the tubes and the spacing between them, do not capture as much of the incident sun as does a flat plate collector. The performance is less affected by small deviations from normal in the incident angle, however, so that overall performance is quite good. Since trough-type reflectors are often used behind the evacuated tubes to improve the proportion of incident solar energy captured, evacuated tube collectors are making some use of concentration techniques. A complete discussion of evacuated tube collectors requires methodology that goes beyond the techniques developed for the analysis of flat plates, so it is difficult to generalize about them. However, several observations can be made. Convection and conduction losses are reduced to near zero in an evacuated tube collector so the importance of a selective coating to reduce radiation loss is more heavily emphasized. Also, the vacuum environment may provide important constraints, both restrictive and beneficial, to the selective coating options to be considered. Needless to say, the evacuated tube collector is an area of active research and development interest in the solar industry.

COLLECTOR HEAT LOSSES

INSULATION AND BACK LOSSES

Heat loss from the back side of a flat plate collector is minimized with thermal insulation. The utility of various insulation materials can be characterized by comparing their thermal conductivity factors, k, which represent the heat loss in BTU per hour per square foot of insulated area per inch of thickness per F degree of temperature difference. Precise calculations of heat loss from the back are used to quantify the characteristics of a specific system but, more importantly, provide another measure of evaluating collectors through the relationship:

$$J_{\text{back}} = \frac{T_p - T_a}{\frac{x}{k} + \frac{1}{h_b}} \qquad\qquad (4\text{--}8)$$

where: J_{back} is the heat loss in BTU/hr-ft² from the back of the collector

k is the insulation k-factor in BTU/hr-ft²-°F-in

x is the thickness of the insulation in inches

T_p is the temperature of the absorber plate in °F

T_a is the outdoor ambient temperature in °F

h_b is the convection coefficient (2 to 4 BTU/hr-ft²-°F)

Equation 4–8 ignores the effect of an insulation box or back pan. In order to include the effects of additional insulation and other materials with different k-factors, it is convenient to use the R-factor where:

$$R = \frac{x}{k} \qquad\qquad (4\text{--}9)$$

and the equivalent values of R for two separate isulation layers can be added.

EDGE LOSSES

Edge losses may be treated in a manner similar to back losses by defining the edge area as the product of the perimeter of the collector and its thickness. If the edge is insulated with material having thermal conductivity, k, and thickness, x_e, the edge loss will be approximated by:

$$J_{\text{edge}} = \frac{k}{x_e} (T_F - T_a) \frac{\text{Edge area}}{\text{Collector area}} \qquad\qquad (4\text{--}10)$$

where: J_{edge} is the edge loss in BTU/hr-ft²-°F

x_e is the thickness of the edge insulation in inches

k is the thermal conductivity of the edge insulation in BTU-in/hr-ft² °F

T_F is the temperature of the absorber plate, °F

T_a is the ambient temperature, °F.

The total losses of a flat plate collector may be represented by:

$$J_T = J_{up} + J_{back} + J_{edge} \qquad \text{BTU/hr-ft}^2 \qquad \text{(4–11)}$$

where J_{up} is the upward loss through the cover plates given in Equation 4–3, J_{back} is the loss through the back of the collector given in Equation 4–8, and J_{edge} is the loss through the edges of the collector given in Equation 4–10.

The thermal loss coefficient of the collector is given by:

$$U_L = \frac{J_T}{T_i - T_a} \qquad \text{BTU/hr-ft}^2\text{-°F} \qquad \text{(4–12)}$$

where J_T is the total loss given by Equation 4–11, T_i is the inlet temperature in °F and T_a is the ambient temperature in °F. This thermal loss coefficient is an important quantity in evaluating the performance of a flat plate collector and should, ideally, be as small as possible.

As can be seen from Equation 4–3, U_L is not a constant, but varies in a complex way with the level of absolute temperature differences. For a given temperature difference the loss term is a constant, and varying the insolation level will yield a straight line plot of efficiency versus $\Delta T/I$. Since under clear sky conditions the level of solar insolation does vary significantly as the sun transverses the sky (its radiation striking collectors at a succession of off angles), the incident angle modifier should be included in calculations in a manner which is treated in the next chapter.

SELECTIVE COATINGS

As incident solar energy is absorbed by the collector, the plate temperature will rise to an equilibrium at that point where the sum of energy delivered and energy lost is equal to energy absorbed. High collector plate temperatures are desirable but, since all three types of heat losses increase as plate temperature rises, with a consequent decrease in energy delivered and system efficiency, strict control of losses are necessary. As has been shown, conduction and convection losses are countered with multiple covers and back insulation, leaving losses by radiation to assume an increasing importance. Radiation losses increase as the fourth power of the absolute temperatures (see Equation A–2) with the maximum losses occurring in the wavelength region given by the Wein Displacement Law (Equation A–4). Wavelengths in the thermal region are significantly longer than the average of wavelengths of the solar radiation region.

If convection and conduction losses were completely eliminated by surrounding the collector with a perfect vacuum, and if the solar constant was absorbed perfectly by the collector plate (as in a blackbody in space), the highest equilibrium temperature is a function of the thermal emissivity. The relationship is expressed:

$$T_{eq} = \sqrt[4]{\frac{I_0}{\sigma\epsilon}} \qquad \text{(4–13)}$$

where: T_{eq} is the highest equilibrium temperature (°K)

I_0 is the solar constant (see Table 2–1)

σ is the Stefan-Boltzmann constant, Equation A–2

ϵ is the emissivity of the absorber at wavelengths near those given by Equation A–4.

For $\epsilon = 1$, such as is defined for a blackbody, $T_{eq} = 124°C$. For most real objects the thermal emissivity ranges from 0.6 to 0.9 and in that range T_{eq} can range from approximately 170°C to 130°C, respectively. Table 4–2 shows the highest equilibrium temperature, T_{eq} in °F for various values of emissivity, ϵ, under the assumptions that: no other losses are present, there is perfect absorptance, and insolation occurs at the solar constant.

It is apparent from the data that the optimum highest possible equilibrium temperature results when collectors are constructed of materials with the lowest thermal emissivity. Unfortunately, most materials with low thermal emissivity are also low in solar absorptance with the effect that equilibrium temperatures above 500°F have not been achieved with flat plate collectors. The quandary is, however, not insoluble.

A selective surface is a coating or surface preparation, black in appearance since the sensitivity of the human eye is limited to that part of the solar spectrum where the surface is an absorber, which combines low thermal emissivity with high absorptance. Many coatings provide high absorptance, but to simultaneously gain low thermal emissivity it may be necessary for the coating to have high thermal and electrical conductance as well. The emissivity of a material is related to its electrical resistivity by Drude's relation:

$$\epsilon = 0.365 \sqrt{\frac{\rho}{\lambda}} \tag{4–14}$$

TABLE 4–2 Effect of Emissivity on Equilibrium Temperature under Idealized Conditions (in Vacuum with No Covers)

Emissivity, ϵ	Equilibrium Temperature, T_{eq} (°F)
0.05	1,027
0.10	791
0.20	592
0.30	491
0.40	425
0.50	377
0.60	339
0.70	309
0.80	284
0.90	262
1.00	244

where: ϵ is the thermal emissivity

   ρ is the electrical resistivity at wavelength, λ

   λ is the wavelength.

Drude's relation is valid where the surface of the material is smooth and maintains its structural integrity over distances significantly larger than the wavelength, λ. Drude's relation serves to show why most conventional black paints (which are non-conducting and, thus, have high resistivity) are not useful as selective coatings.

Many materials and surface preparation techniques have been explored for possible application as selective surfaces. The synthesis of a selective surface technique that is economical and long-lasting is still the subject of basic research, but practical selectively absorbing surfaces are now being produced.

Polished metals typically absorb very little of the solar radiation that is incident upon them. As conductors of electricity, metals have an abundance of highly mobile surface electrons that quickly respond to incident electromagnetic fields and reflect them away from the metal. At the same time, however, these mobile electrons are unresponsive to the thermal mechanical agitation of their respective metallic atoms and do not emit radiation in response to heat as well as other substances. Thus, most metals have both low absorptivity and low emissivity. The mobility of the electrons on a polished metal surface varies with the frequency (or wavelength) of the incident electromagnetic wave or thermal mechanical agitation, however, so that, although absorptivity and emissivity must be equal at any particular wavelength (Kirchoff's Law), they can vary considerably between the short wavelengths of solar energy and the longer wavelengths of thermal mechanical agitation. Table D–5 shows the solar absorptivity and thermal emissivity of commonly used metal surfaces.

Iron and zinc have both been considered as naturally selective materials. Since galvanized steel, which consists of a zinc coating on steel, is inexpensive it seems an attractive possibility. Its solar absorptivity, however, is only 0.5 which, while higher than other metals, is only slightly more than half as absorptive as black paint. Consequently, some experimentation is being conducted on galvanized sheet steel that has been folded accordion-fashion into V grooves where sunlight is forced to reflect several times with successive reductions in net reflectance. The result is some improvement in absorptivity but at the expense of an increase in emissivity resulting from a larger emission surface area for thermal radiation.

Another approach to the synthesis of a selective surface is to coat a metal surface with tiny conducting spheres or holes that absorb light but are smaller than the wavelengths that are characteristic of thermal radiation. Because of this, the tiny black spheres are transparent to thermal radiation and retain the low emissivity of the underlying metal surface. This is the concept used in applying very thin layers of paint containing carbon black, miniscule agglomerated metal droplets, or other optically absorbing ingredients to a metal backing. Difficulties arise in finding a suitable matrix or binder for the spheres, and in obtaining a uniform particle size and spacing, and in providing a uniform thickness.

One of the most successful methods to date of producing selective surfaces involves the use of semiconducting material. Some semiconductors are similar

TABLE 4–3 Practical Selective Surface Materials

Material	Developer	Solar Absorptance	Thermal Emittance
Semiconductor Lead Dioxide (on copper)[1]	AMETEK	0.99	0.25
Black Chromium on Bright Nickel (on aluminum or steel)[2]	Olympic Honeywell NASA-Lewis	0.95 0.977 0.93	0.10 0.19 0.12
Black Nickel[2] (on nickel or copper)[3] (on galvanized iron)[3]	Honeywell Tabor Tabor	0.90 0.90 0.89	0.08 0.05 0.12
Copper Oxide (on copper)[3]		0.89	0.17

[1] U.S. Patent #3,958,554, Dr. Ferenc Schmidt; assigned to AMETEK, Inc.
[2] R. L. Lincoln, D. K. Deardorff and R. Blickensderfer. "Development of ZrOxNy Films for Solar Absorbers," Society of Photooptical Instrumentation Engineers. Vol. 68:1975, page 161.
[3] H. Tabor. "Selective Surfaces for Solar Collectors," Chapter IV of *Low Temperature Engineering Application of Solar Energy,* published by ASHRAE.

to metals in thermal emission and are similar to non-metals in their ability to absorb solar energy. The semiconducting oxides of iron, cobalt, vanadium, nickel, chromium, copper, and lead all have selective properties and have been shown to perform as selective surfaces. Table 4–3 lists the properties of some of the selective coatings that are practical for use and commercially available.

The selective surface materials listed in Table 4–3 are limited to surfaces that can be prepared by electrodeposition with relatively cost-effective raw materials. There are hundreds of techniques of producing selective surfaces by vacuum deposition but the process is expensive when large, planar areas must be covered. Meinel and Meinel (*Applied Solar Energy,* Chapter 9. Addison Wesley: 1976) have discussed many selective surfaces which include those requiring vacuum deposition and those which use multiple layer interference stacks which operate on principles similar to those of optical antireflection coatings. These more advanced types of selective coatings are not cost effective for flat plate collectors which require large planar areas, but they may have significant application in concentrating collectors where the active absorber surface has considerably less area than the solar interception aperture area.

ENCLOSURE AND SUPPORT FOR SOLAR COLLECTORS

The cover assembly and absorber plate require support; ideally, the absorber plate should be supported independently of any thermally conducting outer frame to limit heat losses by conduction and to allow for thermal expansion. If multiple glazings are used, the inner glazing should also be supported independently since its temperature will rise, but the outer cover can be supported by the external enclosure or frame since it is, itself, in contact with the outside air.

Alternative methods of utilizing an extruded frame and separate insulation backing pan may be acceptable in situations where high efficiency is not

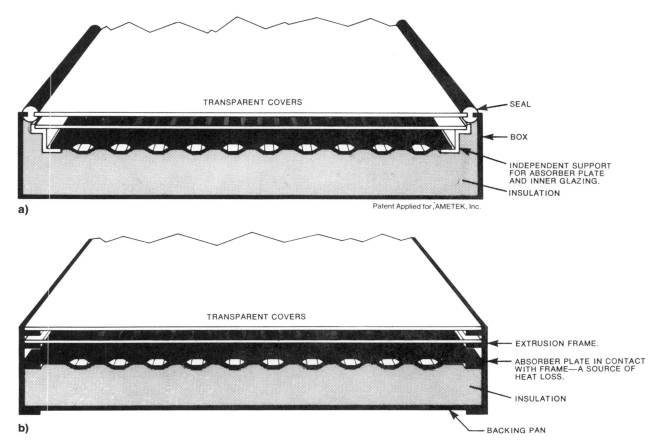

TRANSPARENT COVERS

SEAL

BOX

INDEPENDENT SUPPORT
FOR ABSORBER PLATE
AND INNER GLAZING.

INSULATION

a)

Patent Applied for AMETEK, Inc.

TRANSPARENT COVERS

EXTRUSION FRAME.

ABSORBER PLATE IN CONTACT
WITH FRAME—A SOURCE OF
HEAT LOSS.

INSULATION

BACKING PAN

b)

FIGURE 4–9   Two alternative enclosure designs.

needed, air leakage can be tolerated, and structural support is provided externally to the collector itself. But a completely enclosed, structurally rigid, airtight enclosure provides the best support for a flat plate collector. This airtight enclosure should be provided with a breathing tube which allows for thermal expansion and contraction of air through a desiccant substance; the desiccant controls the moisture content of the air within the collector since moisture condensed on the inner glass cover can reduce the efficiency of a collector, as well as impair the properties of some selective surfaces.

Figure 4–9 diagrammatically compares two enclosure designs. Drawing (a) used a rigid enclosure, independent support for the absorber plate and inner glass cover. The design in (b), while it offers some advantages in cost, is subject to greater edge losses.

CONCENTRATING COLLECTORS

Concentrating collectors use mirrors or lenses to focus light from a large area onto a small area where the more intense light can achieve higher temperatures and a greater conversion into usable energy with less receiving area.

93

Concentration with mirrors is possible in several ways. To focus light onto a small area, conical, spherical, or paraboloidal mirrors are used. Linear, circular, or parabolic reflectors will focus light onto a narrow line. A large number of flat mirrors (or a Fresnel reflector) can be used to concentrate light onto a single place or a narrow line.

Lenses can also be used for light concentration. A plano-convex lens, similar to the familiar magnifying glass, will focus light onto a small area. Convex, cylindrical lenses will yield a narrow line. A Fresnel lens uses adjacent flat refractors to accomplish the same effect but can be made in a flat sheet rather than a convex shape.

No matter what the means employed, concentration is a matter of degree which is specified by the concentration ratio (CR); the concentration ratio of a concentrating collector is the ratio of area of sunlight intercepted to the area onto which it is concentrated. Figure 4–10 depicts the effect of concentration ratio, CR, on the equilibrium temperature of an object placed at the focus of a concentrating collector. The lower limit shown represents the concentration ratio required to offset thermal losses. Any useful gain would require the higher CR shown in the shaded area.

Concentrating schemes can only make use of the direct beam contribution of total solar insolation so the total energy collected will be less than that collected by a flat plate intercepting the same area, since the flat plate collects both direct and diffuse radiation. Thus, the principal reason for using concentration is to increase the operating temperature. It has been argued that, given a stated interception area of solar insolation, a reflecting concentrating collector may be more cost effective than a flat plate absorbing area, but the cost advantage is usually more than equalled by higher initial, maintenance and operating costs. Moreover, dirt and scratches are more degrading for concentrating collectors. While all these considerations result in fewer applications, concentrating collectors still comprise the second largest group of solar energy systems, with a number of variations worth noting.

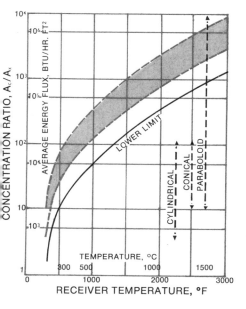

FIGURE 4–10 The effect of concentration ratio on receiver temperature.

MULTIPLE HELIOSTAT CONCENTRATORS

The first concentrating technique, of use primarily in large, central utility installations, is the concentration of sunlight by reflecting the solar energy from a number of flat mirrors onto a central receiver. Each of the flat mirrors must be individually tilted on two axes with a central computer to provide the necessary tracking information. These rotating mirrors are called "heliostats" and a multiple heliostat concentrator is shown diagrammatically in Figure 4–11. The degree of concentration attained in a multiple heliostat concentrator is large enough to provide temperatures necessary for the generation of electricity from super-heated steam. Thus, although it is suitable for solar energy utilization in utility power plants, it is not considered appropriate for supplying solar energy directly to individual buildings and houses.

The technologies employed in the central receiver can be more advanced because its cost is based on the large area of sunlight intercepted by the heliostats rather than upon the small area of the receiver itself.

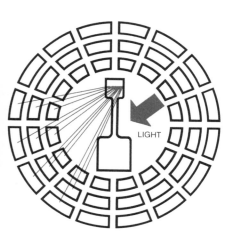

FIGURE 4–11 Fresnel reflector: a multiple heliostat concentrator.

FOCUSING REFLECTORS

Reflecting bowl-shaped surfaces may be used to concentrate solar energy onto an area smaller than the area of sunlight intercepted by the bowl. The highest concentration ratio is obtained with a paraboloidal surface that focuses sunlight onto a small disc representing the optical image of the sun in a manner similar to that employed in reflecting telescopes. In fact, any portion of such a surface, on or off its principal axis, can serve as a focusing device. Like each of the mirrors in a multiple heliostat concentrator, a paraboloid must be rotated on two axes to follow the sun. With one axis properly aligned in parallel to the earth's axis of rotation, the tracking during any particular day can be achieved on one axis while seasonal tracking is obtained on the other. A paraboloidal reflector is shown diagrammatically in Figure 4–12.

A reflecting spherical bowl can also be used to concentrate sunlight. Since the requirement for a well-defined solar image is not fully essential for most concentrating applications, it is possible to sacrifice a small amount of concentration ratio to gain other benefits. Because a spherical surface is symmetrical to incident angles of light, tracking of the sun does not require rotation of the entire surface, only the receiver. This principle may be applied where very large stationary bowls are fabricated by mounting reflectors into extinct volcano calderas, meteor craters, or other depressions in the ground with a suspended, mobile, central receiver to track the moving focal region of the sun's energy. The same technique is used in some radar antennae and large radio telescopes.

A reflecting conical bowl will concentrate light also, but to a lesser degree than a spherical or paraboloidal reflector. Its primary advantage is that it can be fabricated from a flat sheet of reflecting material because it is a simple curved surface, in contrast to the others which are compound curved surfaces.

Focusing collectors can also be made from troughed surfaces that focus sunlight onto a long pipe or linear region, as shown diagrammatically in Figure 4–13. The concentration ratio is fundamentally smaller than that for bowl collectors but this trough type does have many practical advantages. The reflecting surfaces, whether of parabolic, circular, or V cross section, are simple curved surfaces capable of fabrication from flat sheets. The focal region of a trough is well suited to the transfer of heat to conventional fluid pipes and the tracking requirements along at least one of the two axes are less stringent if the orientation of the collector is properly chosen.

LENSES AND REFRACTION CONCENTRATORS

The refracting property of transparent materials can be utilized by constructing lenses of glass or any of a wide variety of transparent plastics to concentrate light onto small focal regions. Three of the most commonly seen lens geometries are diagrammatically shown in Figures 4–14, 4–15, and 4–16. Just as with reflectors, refracting devices can be designed to concentrate light onto a region surrounding a single point or onto a line conforming to the geometry of fluid pipes. The constraints for tracking are also quite similar to their geometric counterparts among reflectors.

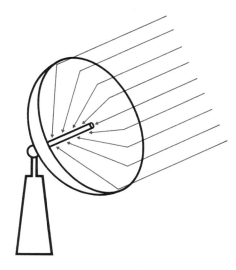

FIGURE 4–12 Paraboloidal reflector.

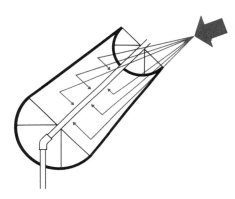

FIGURE 4–13 Cylindrical parabola reflector.

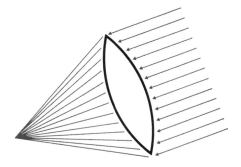

FIGURE 4–14 Convex lens (edge view).

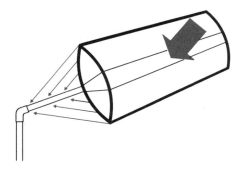

FIGURE 4–15   Linear convex lens.

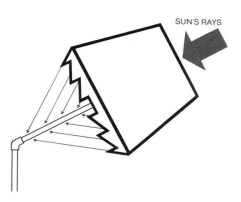

FIGURE 4–16   Linear Fresnel lens.

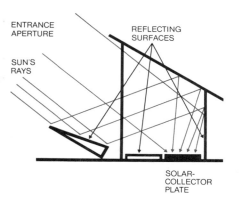

FIGURE 4–17   Pyramidal reflector.

Refracting elements offer some advantages over reflecting techniques since the refracting element can serve the dual purpose of acting as a convection suppressing cover plate, as well as a concentrating element. Under the most practical conditions of using the most economical materials, the transmittance of most refracting elements is greater than the reflectance of reflecting surfaces. Thus, refracting concentrators can usually collect more energy per area of sunlight intercepted than would reflecting techniques.

FIXED CONCENTRATORS

Because it is widely acknowledged that one of the principal constraints in using concentrating collectors is the necessity for tracking the sun, a great deal of attention and effort has been directed toward the development of a fixed concentrator. Fixed concentrators are usually designed to operate within a prescribed acceptance angle over which the concentrator can collect sunlight.

The acceptance angle of a fixed concentrator is dependent upon the concentration ratio in a manner which is a function of the geometry of the collector. Invariably, the acceptance angle tends to increase as concentration ratio is decreased. Ideal reflector geometries which optimize the combination of acceptance angle and concentration ratio have been described, particularly by Winston, so collectors of this type are often called Winston Concentrators or compound parabolic concentrators (CPCs).

Another class of fixed concentrators is known as pyramidal concentrators. These make use of multiple reflections into more confined areas where the solar collection element is located. A typical example of a pyramidal reflector for a fixed concentrator is shown in Figure 4–17. Among the limitations of these reflectors (which use flat reflecting surfaces in the main) is the necessity for multiple reflections which successively reduce the incident energy intensity; a second drawback is the large reflector area compared to collection aperture area. In addition, this type of collector is vulnerable to the effects of wind and snow.

CPCs and other ideal geometries require curved surfaces that may be difficult to fabricate and support. Many of the CPC geometries, including the more exotic "seashell" curves, toroidal and involuted shapes, despite their undeniable elegance, require large surface areas of reflector as compared to the aperture area of sunlight collected.

TRACKING TECHNIQUES

Concentrating collectors, for the most part, must follow the sun as it moves across the sky to gain efficiencies which justify their added costs. The accuracy with which a concentrating collector must follow the sun increases with the concentration ratio, CR. As has been shown, some collector geometrics, particularly those with spherical or cylindrical reflectors, allow the reflector surface to remain fixed while the receiver is tracked. It is also possible to rotate a flat reflecting surface around the earth's polar axis and reflect the sun's rays in a fixed direction along the polar axis so that concentration can be

96

achieved without tracking any of the collector components. Figure 4–18 shows a few of the techniques used for tracking in order of their increasing accuracy.

The action of sunlight on a temperature sensitive bimetallic helix will provide sufficient tracking force for the lightweight thermal heliotrope shown in Figure 4–18*a*. As the collector approaches its correct position, the shade automatically regulates the amount of sunlight heating the coil, a continual process which allows the collector to progressively advance in concert with the sun.

Paired photocells (Figure 4–18*b*) are used to provide an electrical signal which is a function of the degree of tilt of the collector surface from the normal to the sun's direction. Motor drives are set to respond to the signal by tilting the surface in the same direction that reduces the signal, thus following the sun.

Tracking on two axes can be achieved by the use of a pinhole mounted over a photocell array, as shown in Figure 4–18*c*. Two motors are operated

FIGURE 4–18 Tracking techniques.

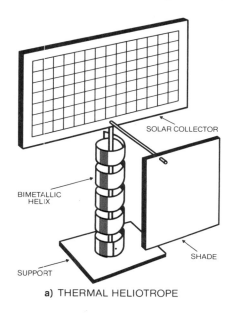

a) THERMAL HELIOTROPE

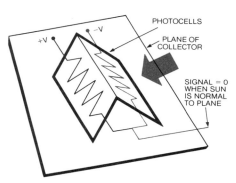

SIGNAL IS USED TO CONTROL MOTOR
MAT TILTS PLANE UNTIL SIGNAL IS ZERO
b) PAIRED PHOTOCELLS

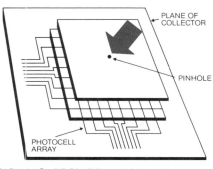

SIGNALS PROVIDE INFORMATION ON
SUN'S POSITION; MOTORS CAN
CHANGE POSITION IN TWO DIMENSIONS.
c) PHOTOCELL ARRAY

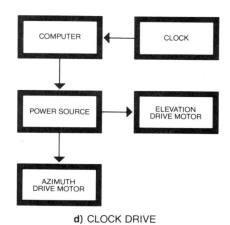

d) CLOCK DRIVE

by signals from the opposing pairs of photodetectors. In the same manner, an array of photocells under a pinhole can be used to maintain an adjustable tilt angle.

Clock drives (Figure 4–18*d*), operating from the precise determinations of the sun's position made possible with computers, are used in tracking large or heavy collectors in the same manner as their use in telescope tracking. Unlike other tracking techniques, a clock drive can operate when the sun is obscured and, indeed, will track any object in the sky, visible or invisible, which follows a known or predictable path. It is, therefore, the most accurate technique.

PHOTOVOLTAIC COLLECTORS

Because electricity is one of the most versatile forms of power, the ability to directly convert solar energy into electricity is of great interest. Among the methods that have been devised for this conversion, the most successful have been those which apply the *photoelectric effect*. Described on the atomic level, the photoelectric effect is the excitation of a conducting electron from a stable orbit by the direct action of light of sufficiently high energy. A portion of the spectrum of sunlight is capable of energizing electrons to forceful conduction as it interacts with certain materials, especially those junctions between oppositely doped semiconductors.

There are other phenomena that can result in the production of an electric current from the action of light. When two dissimilar metals are joined at a junction which is heated by the action of light, an electric potential results from the *thermoelectric effect;* when light of sufficient intensity strikes any metal which is sufficiently heated in a vacuum, electrons will be released by the *thermionic effect*. Both phenomena, although they have been investigated for solar applications, are one step removed from direct conversion to electricity, since they depend upon the conversion of sunlight to heat before producing electricity. Thus, they tend to be less efficient than the photoelectric effect in producing electricity. In addition, there are other methods involving the differences in ionization potentials characteristic of various phases of some materials or, most common of all, the conversion of mechanical energy into electricity through an engine or turbine driven generator. In large scale solar applications, especially those using concentration techniques, there is no *a priori* reason to expect that photovoltaic converters based on the photoelectric effect will soon be more cost effective in producing electricity than the steam cycle.

Photovoltaic cells do offer some unique advantages. They have no moving parts which, in principle, translates into less maintenance; they may be used advantageously with concentration but do not require it; and, since operation does not involve the use of heated fluids or other materials, insulation is unnecessary. Further, the flexibilities of transmission and storage techniques are greater for electric power than for thermal power, photovoltaic cells are light in weight with greater portability, and they may be used in very small areas, as in powering wristwatches.

Photovoltaic cells are made from several materials, with silicon and cadmium sulfide the most commonly used in commercial units. Table 4–4 lists the range of efficiencies available from many of the materials which have been investigated.

TABLE 4-4 Photovoltaic Solar Cells

Material	Efficiencies (%)	Availability
Silicon wafer	12–18	Commercial
Silicon, thin film	2–5	Experimental
Gallium arsenide	16–20	Experimental
Cadmium sulfide	5–8	Commercial
Cadmium telluride	5–6	Experimental
Silicon carbide	1–3	Experimental
Gallium phosphide	1–3	Experimental
Indium phosphide	2–5	Experimental
Zinc selenide	NA	Experimental

The diagrammatic description of the operation of a silicon photovoltaic cell in Figure 4–19 illustrates some of the sources of inefficiency when various wavelengths of sunlight interact with the silicon material.

The power that can be obtained from a photovoltaic cell is the product of the voltage, v, and the current, i, shown applied to a load resistance, R, in Figure 4–19. Maximum power is obtained when the load resistance, R, is matched to the ratio of v to i at the point of maximum power ($v \times i$) in the load

FIGURE 4-19 Diagrammatic representation of the operation of a typical silicon solar cell.

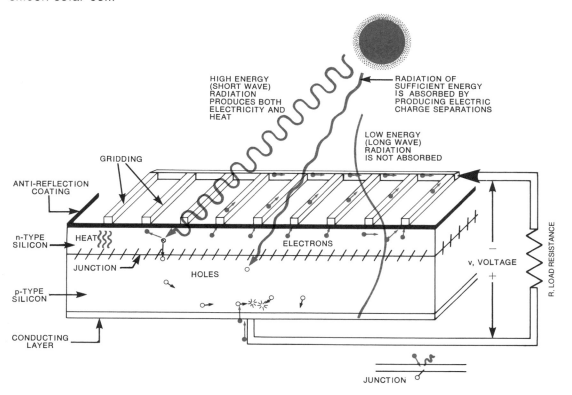

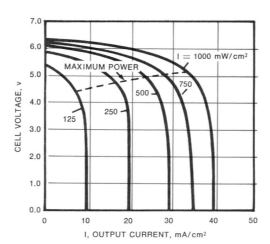

FIGURE 4–20 Characteristic curves for a silicon solar cell. Variations in solar insolation, *I*, produce a consequent variation in matched load resistance, *R*.

characteristic curves shown in Figure 4–20. Note that the material value of load resistance increases with reductions in solar insolation in order to remain on the maximum power line under all conditions. The characteristic curve also changes with temperature and is dependent upon the method of gridding used to collect the electric current from the cell.

The efficiencies shown in Table 4–4 assume that the maximum power is being delivered by the solar cell to a matched load. Since the characteristic curves indicate that the matched load required for maximum power output changes with solar insolation and temperature, it is not as easy as it may seem at first sight to realize these efficiencies in practice without sophisticated controls.

The principal impediment at the present time to the widespread use of photovoltaic cells is high costs which are substantially higher than the level at which power generation under peak conditions would be competitive with existing conventional electric power rates. Although many important developments have contributed to successive cost reductions in the recent past, there is no certainty that solar cells will soon become competitive. The proponents of solar cells extrapolate future cost reductions from past efforts based upon anticipated economies of scale and programmed technical development efforts, but its role will probably be confined to specialty applications for many years to come.

CHAPTER 5
COLLECTOR EFFICIENCY

The efficiency of a flat plate solar collector is an important parameter for determining the collector area required to provide a needed amount of solar thermal energy and, therefore, a measure to be used in conjunction with other properties (shape, weight, strength, durability, and orientation, to name a few) as a means of comparison between various collector designs. While efficiency is a vital concern, it is also one of the most difficult of the properties of a collector to measure accurately and is potentially subject to misinterpretation and abuse since it is neither an obvious property nor a single quantity. Furthermore, the value of efficiency that pertains to conditions at the solar noon hour is quite different from the efficiency that applies for the course of a day, month, or year. This is because the instantaneous efficiency changes with different levels of solar insolation in a manner prescribed by the curves shown in Figure 4–7. Accordingly, before progressing to the mathematical relationship which defines efficiency, it is of value to first review the sources of potential collector losses with a brief note of the countermeasures specified in Chapter 4. Also relevant is the consideration of the accuracy of the means of measurement for all collector properties which is discussed in Appendix C.

COLLECTOR LOSS SUMMARY

Fin conduction loss due to thermal gradients in the absorber plate occurs when heat generated by the absorbance of sunlight in the absorber plate is not transferred into the fluid. This loss results when portions of the absorber plate are not in close proximity to the fluid conduits, when the thermal conductivity of the absorber plate is poor, or when the fluid flow is poorly distributed. Fin conduction loss can be minimized by the proper choice of plate materials, construction method, and fluid flow pattern in the absorber plate.

Thermal radiation loss results from the net electromagnetic radiation which is emitted from any part of the collector that is heated. This loss tends to increase as the fourth power of the absolute temperature rise in accordance with the Stefan-Boltzmann equation and can be reduced by using selective coatings of low emissivity and high absorptance for solar radiation on the absorber plate. Another technique is to use selectively reflecting "heat mirrors" that reflect the emitted energy back onto the absorber plate without disturbing the incoming solar energy.

Convection losses are due to the effects of air circulation between surfaces of different temperatures. As the absorber plate temperature rises, convective movements of air carry its heat onto adjacent cooler surfaces. Suppression is usually accomplished by using multiple covers or honeycomb suppressors that reduce the freedom of air to move without disturbing the incoming solar radiation.

Conduction losses occur through materials that are in direct contact with the absorber plate. Such losses can be reduced by insulating the back and edges of the absorber plate with materials which have low thermal conductivity.

Losses due to condensation and air leakage in the collector enclosure are minimized by using airtight construction for the enclosure and by controlling the humidity of air that must enter and leave the collector as its temperature rises.

COLLECTOR MEASUREMENT

The measurements needed to determine the instantaneous efficiency of a flat plate collector through Equation 5–1 are temperature, fluid mass flow, solar insolation and area. Temperature measurements are required at the inlet and outlet of the collector with the difference between the two, ΔT, often measured independently to improve upon the accuracy of the measurement. The ambient temperature surrounding the collector and the total intensity of solar radiation, including direct and indirect components incident to the collector aperture and in the plane of the collector, must be measured. Finally, the rate of mass fluid flow through the collector and the area of the collector aperture must also be measured in order to determine the instantaneous efficiency.

These measurements are the terms of the instantaneous efficiency equation, but the implementation of a test procedure for purposes of comparing collectors under equivalent conditions implicitly requires that many other quantities be known as well. For example, the tilt angles of the collector in its test setup, the specific heat of the liquid used in the collector, the wind speed and direction, and solar time should be known. Errors in the determination of these quantities can produce errors in efficiency measurement, but most experimenters have found that such errors are either very small, have little effect on the end result, or need be established only infrequently. Fortunately, both the National Bureau of Standards (NBSIR 74-635) and ASHRAE (93-77) have established procedures for efficiency measurement test conditions; these procedures provide the standard conditions used by experienced independent testing laboratories in implementing test procedures. Accuracy in measurement of these quantities is most critical to those engaged in that pursuit—manufacturers and independent testing laboratories performing comparative testing.

The same limitations on accuracy inherent in the instruments and methods of measurement which impact on comparative testing can affect accuracy in instantaneous efficiency calculations for collectors at a given site. A complete discussion of limiting accuracies of measurements for temperatures, liquid mass flow, solar insolation, and collector area, as well as a discussion of accuracy, tolerance, precision, and sensitivity in any series of measurements, is contained in Appendix C.

The instantaneous efficiency of a flat plate collector is defined as:

$$\eta = \frac{\dot{m} C_p \Delta T}{A_c I_T} \qquad (5\text{--}1)$$

where: η is the efficiency of the collector expressed as a number between 0 and 1

$\dot{m}$ is the rate of fluid mass flow through the collector (lb/hr)

C_p is the heat capacity of the liquid (BTU/lb-°F, water = 1.0)

A_c is the gross area of the collector (ft²)

I_T is the total solar insolation (BTU/hr-ft²)

ΔT is the temperature difference between the outlet and inlet fluid $\Delta T = T_0 - T_i$ (°F).

As the average fluid temperature of the collector increases, thermal losses increase and the efficiency decreases. Figure 4–7 shows a plot of the efficiency of a typical flat plate collector with one, two, and three transparent covers, and with or without a selective coating. The importance of the selective coating increases with higher average fluid temperature, T_F, as:

$$T_F = \frac{T_{\text{out}} + T_{\text{in}}}{2} \qquad (5\text{--}2)$$

where T_{out} and T_{in} are the outlet and inlet fluid temperatures, respectively.

The efficiency of a flat plate collector may also be expressed in terms of the solar energy, I, intercepted and the fundamental quantities related to the plate's construction. Following Duffie and Beckman:

$$\eta = F_R A_x \left[K(\tau\alpha)_e - \frac{U_L(T_i - T_a)}{I} \right] \qquad (5\text{--}3)$$

where:

$$F_R = \frac{\dot{m} C_p}{U_L} \left[1 - e^{\frac{-U_L F'}{\dot{m} C_p}} \right] \qquad (5\text{--}4)$$

$$F' = \frac{1}{w \left[\dfrac{1}{F(w - u) + u} + \dfrac{U_L}{h_o \pi (u - 2t)} + \dfrac{U_L}{Cb} \right]} \qquad (5\text{--}5)$$

and where:

η = thermal efficiency of the collector

F_R = overall efficiency factor

F' = collector geometry efficiency factor

F = fin efficiency factor (Equation 4–2)

A_x = A_p/A_g

A_p = area of the absorber plate

A_g = gross area

$K(\tau\alpha)_e$ = previously defined (Equation 4–5)

U_L = previously defined

$\dot{m}$ = previously defined

C_p = previously defined

w = distance between tubes

u = outside diameter of circular risers in the plate

t = tube wall thickness

h_o = convection coefficient between the tube wall and circulating fluid

C_b = bond conductance between fin and riser where

$$C_b = \frac{k_b b}{\gamma}$$

k_b = thermal conductivity of the material bonding the plate and fin

b = bond length

γ = bond thickness

The equation for efficiency, 5–3, is divided into two terms within the brackets. The first term describes the solar insolation which is absorbed, a quantity which is obviously influenced by the intensity and angle of incidence of solar radiation incident on the plate. Insolation is reduced by incidence at angles off direct beam incidence chiefly by the cosine law (Equation 2–12), increased reflection from the outermost glazing (Equation 4–5), and shadows cast upon the plate by the edges of the enclosure (Equation 4–17.) These off-angle reductions act only on the first term in the efficiency equation.

The second factor in Equation 5–3 describes losses. As noted in Equations 4–3 and 4–12, the loss term is a complex quantity which is not a linear function of the temperature difference between the inlet temperature to the collector and the ambient temperature. For any *given* temperature difference between the plate and ambient, the loss term is a constant. The elevated temperature off-angle performance will be along a series of lines parallel to the slope of the noon curve at the point defined by the inlet and ambient temperatures.

Off-angle performance is dependent on the materials used to construct the solar device, with each of the material parameters affecting the rate of change of U_L with T. A single glazed collector with a high emissivity coating is most affected, low emissivity coatings the least, with extra glazings reducing the rate of change. The calculated constant temperature off-angle performance of two different coatings on an efficient absorber plate, and operating under similar conditions, are shown in Figures 5–1 and 5–2.

It should be noted that with 50% of clear day summer performance output at 70°F ΔT is within 30° of noon, 82% within 45° of noon, and 96% within 60° on either side. Winter percentages are even more in favor of the angles near noon. This is the foundation for the recommendation made by Duffie and Beckman, and others, to simplify the performance plot by assuming (1) average U_L as a constant; (2) all day performance along this plot of linear equation; (3) a correction factor to the intercept of 0.93 for single glazed collectors and 0.91 for double glazed collectors to bring all day prediction in line.

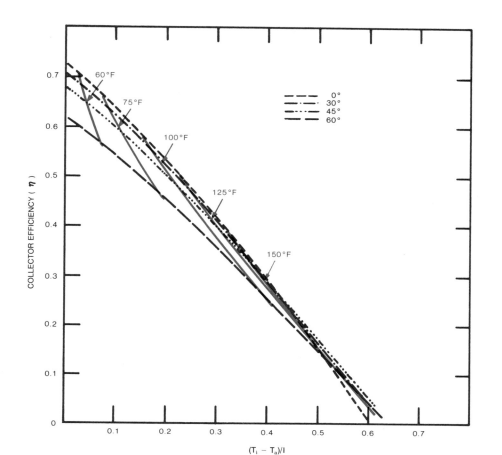

FIGURE 5-1 Off-angle performance curves for constant inlet temperature using an absorber with a flat black painted surface.

FIXED PARAMETERS

$\tau = 0.91$	$I_{30} = 263$
$a = 0.95$	$I_{45} = 200$
$\epsilon = 0.95$	$I_{60} = 120$
$I_0 = 317$ BTU/hr ft²	$v = 5$ mph
Tilt = 40°	$T_a = 50°F$

These assumptions are workable because off-angle performance at any reasonable $\Delta T/I$ is either slightly above, below, or on the linear fit noon curve according to the particular collector being studied.

Several other general observations are noteworthy. At sufficiently low values of operating parameter, ν, the performance is not improved by increasing the number of covers because the effects of the additional loss of solar transmission through the covers are greater than the decrease in the loss coefficient, U_L. Conversely, at high values of operating parameter, the improvement in performance as more covers are added is significant. Improvement always results from the use of a selective absorber coating, provided that its absorp-

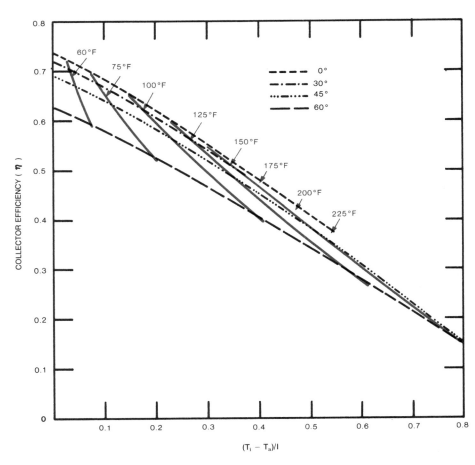

FIGURE 5-2 Off-angle performance curves for constant inlet temperature using an absorber with a black chrome surface.

FIXED PARAMETERS

τ = 0.91	I_{45} = 200
a = 0.95	I_{60} = 120
ϵ = 0.1	v = 5 mph
I_0 = 317 BTU/hr ft²	T_a = 50°F.
I_{30} = 263	Tilt = 40°

tance of solar energy is not less than the non-selective coating. The degree of improvement increases with operating parameter but decreases with an increasing number of covers. Thus, the choice of the number of covers and whether or not to use a selective coating is dependent upon the operating parameter and the costs of various alternatives. Clearly, if plate temperatures in the neighborhood of 200°F are to be achieved, a selective surface is usually well worth its additional cost because of the marked improvement in efficiency.

CHAPTER 6
SYSTEMS FOR HEATING
AND COOLING

Systems to collect, store, and utilize solar energy may vary greatly in design because of the wide choice of equipment, the diversity of design imposed by differing needs at each installation site, and the consideration of costs versus capacity. No matter what choices are made for individual applications, however, every solar thermal system must provide for seven basic requirements: interconnection of collectors, a means of fluid circulation through the collectors, a technique to transfer thermal energy from one fluid to another, a mechanism for energy storage, a means of converting thermal energy to useful forms (such as cooling), a control system, and a technique for delivering the energy in the appropriate form to its intended application. Note that "fluid" is used above in its broadest sense, i.e., both liquids or gases. This chapter, in the main, will address systems which use a liquid as the heat transfer medium because, for reasons noted in the earlier discussion of air-heating collectors, these liquid-heating collectors and the systems which they power are expected to be in most common use.

SERVICE REGIMES

As a first step toward a survey of solar energy equipment, solar energy systems can be categorized through one parameter of their application: the requirement for low, medium, or high temperature service regimes. This division indicates that a system which is cost effective in a certain temperature regime is suited for that particular application only. Indeed, it is not advisable, and often not possible, to use the same system for multiple applications in different temperature regimes without resorting to systems consisting of hybrid collector types and multiple storage units.

The low temperature service regime, ranging from below ambient to about 120°F, is typically used with solar heaters for swimming pools and for horticultural purposes; these simple systems can utilize low-cost collectors with high loss rates U_L. The operating parameter for such systems is usually quite low, in the vicinity of 0.1 or 0.2, so the incremental improvement in the performance of the collector through the addition of glazing covers, selective coatings, antireflection heat mirrors, and other advanced features is usually not worth the added cost. As a result, systems that operate in low temperature regimes may use low-cost black plastic or rubber collectors, no glazing, and seek further economies in other components of the system as well. For example, low-cost materials such as PVC pipe may be used (if building codes permit) since it is not harmed by the low temperature fluids. It is of paramount importance, however, to observe that an efficient system in the low temperature regime may fail completely if operation at higher temperatures is attempted. As with any system, losses will increase rapidly with operating temperature, but this property acts as a safeguard in a low temperature application since it prevents accidental operation at temperatures that may be harmful to the other components of the system.

The medium temperature regime is used in hot water service or space heating applications since the necessary temperature range of 120° to 180°F is appropriate. The design of the collectors used in these systems must be sufficiently improved to reduce the loss rate since reasonable efficiencies will be achieved only when operating parameters fall in the range of 0.2 to 0.5. As shown in Figure 4–7, additional glazings and selective surfaces, although not mandatory, do provide an increment of improvement which is significant. Of course, system components must be able to withstand long exposure to medium temperatures.

The high temperature service regime, above 160°F, is a necessity for air conditioning applications and requires high performance collectors whether flat plate, the FEA classification of "special" evacuated tube collectors, or concentrating collectors. Operating requirements for these systems are much more stringent since, with the operating parameter exceeding 0.5, the use of multiple glazings and selective coatings is not only economically effective but nearly mandatory. The system components must be able to withstand both the effects of boiling temperatures for long periods of time and extreme temperature cycling; greater skill must be exercised in controls, and safeguards against pressurization due to boiling water in stagnation must be considered.

INTERCONNECTING COLLECTORS

In most applications, the necessary collection area is assembled through the interconnection of a number of discrete, modular solar collection panels. The method of interconnection of these panels should recognize three needs: flow should be upward through each collector for best efficiency, especially when parallel banks of collectors are fed from the same header; flow should be uniform in each collector so that no panel will operate at excessive temperature and lose efficiency; and the total length of interconnection piping should be kept to a minimum, a strategy which not only contributes to materials economy but increases system efficiency by keeping to a minimum that heat loss which is proportional to pipe length.

Connecting collectors in series reduces the length of the headers required and, consequently, loss of heat from them. The flow rate through an array of collectors connected in series should be adjusted so that the temperature rise contributed by each collector in the array is equal. The temperature rise per collector can be reduced by increasing the flow rate which slightly increases the heat transfer in the plate. If the flow rate is proportionally increased by the number of collectors in series, the losses will be equal to those of collectors

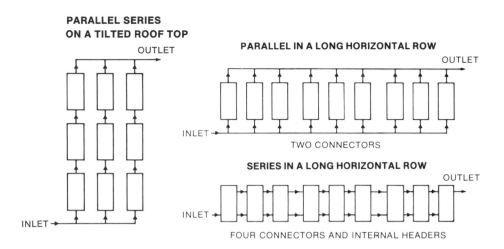

FIGURE 6–1 Acceptable interconnection configurations.

connected in parallel. This is usually only practical with collectors which experienced small pressure drops at the increased flow rate.

If, on the other hand, collectors are connected in parallel, care must be taken to ensure uniform flow in each collector. This is usually accomplished by using low resistance piping for the parallel headers. Figure 6–1 illustrates some of the more economical methods of interconnecting collectors in a bank.

There are two generally relevant considerations that should be observed. The first is that the number of collectors in each series of parallel banks should be equal; collector arrays containing certain large numbers of panels, such as 17 or 37 or any other prime number, are therefore difficult to arrange in convenient, equally sized groups. Secondly, the location of the inlet and outlet fluid connections on the collector will influence the interconnection geometry. Fluid connections on the top and bottom walls facilitate economical header use in the parallel configuration shown in Figure 6–1. Many collectors, to provide convenience in a series parallel configuration, use internal headers that provide fluid connections on the upper and lower ends of both sides of the collector. This method does economize on pipe insulation by providing headers within the collector enclosure, but it also has the dual disadvantages of requiring four connections for each collector instead of two, and more difficulty in maintenance of the internal headers.

FLUID CIRCULATION TECHNIQUES

Solar collectors which use a liquid for heat collection may be either of the open channel or closed tube configuration. The open channel type requires a circulation pump to move the required amount of fluid and to provide the head difference between the system's highest point and its lowest. Circulation in a closed tube system is just the opposite, with the inlet at the bottom and outlet at the top, so that the closed tube may be operated either as a thermosiphon or by forced circulation. In either case, the efficiency of the closed tube configuration is greater than that of the open channel as a result of one effect of siphoning on a closed system which eliminates the need for the circulation pump to work against the collector head difference. Efficiency would be reduced if a closed tube system were connected in the same manner as an open channel, because of the greater work required of the circulation pump and the upside-down temperature distribution on the collector absorber plate.

Fluid circulation in a closed tube system can be accomplished without a circulation pump by using the thermosiphon effect, i.e., the reduction in density of a fluid as it is heated. Thermosiphon systems can be utilized only in applications where the collector is located below the storage or service site, since the fluid must rise from the collector to storage or use. Figure 6–2 illustrates the proper orientation of a thermosiphon system with the storage tank inlet on the top and the outlet at the bottom.

The rate of flow in a thermosiphon system depends upon the size of the tubing used, the vertical height separating the collector from the storage tank, and the temperature difference between the collector fluid and the storage fluid. The flow rate is not necessarily that which can deliver the greatest thermal energy. The use of a forced circulation pump in the system can increase the rate of energy flow to the storage tank, increase the storage tank temperature, and reduce the temperature drop across the collector, as compared to a thermosiphon system operating under the same conditions.

A thermosiphon system requires some means of preventing reverse flow; a check valve is often used. Without such a measure, heat will be removed from the storage tank at night as its fluid reverses through the collector, which then radiates thermal energy to the cold sky. The use of a circulation pump does not obviate the need because many pump designs will allow reverse flow whenever the pump is not operating.

Whether closed conduit or open channel flow is the chosen collector configuration, fluid circulation through the entire solar thermal system may be in either a closed or open loop configuration. The open loop configuration (in which the fluid, in most cases, passes through the system only once) is used in low temperature applications for agriculture and in air-heated ventilator inlets and is generally confined to the use of water or air as the means of heat transfer. The closed loop configuration continuously recirculates the same fluid in the collector; thus, it can use special fluids, achieve higher temperatures, and is not as susceptible to the effects of contaminants in the fluid. The choice between an open or closed loop system will depend on many factors, including the climate, the service application, local water pressure, and building code requirements, but a few general observations are possible.

An open loop system has several advantages over a closed loop in hot water heating service. As in Figure 6–3, water may be taken directly from the mains and preheated by passing through the solar collectors under its own pressure before entering the hot water tank. A circulation pump is needed to maintain circulation through the collectors when hot water is not being drawn. In addition, since potable water is being circulated in the collectors, it is desirable to use a collector with a copper or steel absorber plate to prevent corrosion of dissimilar metals in contact with water. Countering these advantages is the necessity for the collector to be able to withstand street water pressures (unless a pressure reducer is used) and for a freeze drain system in climates where

FIGURE 6–2 A Thermosiphon system.

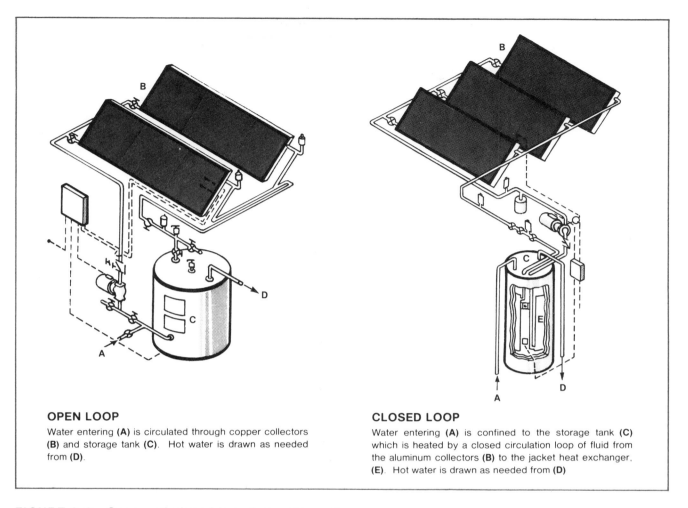

OPEN LOOP

Water entering **(A)** is circulated through copper collectors **(B)** and storage tank **(C)**. Hot water is drawn as needed from **(D)**.

CLOSED LOOP

Water entering **(A)** is confined to the storage tank **(C)** which is heated by a closed circulation loop of fluid from the aluminum collectors **(B)** to the jacket heat exchanger. **(E)**. Hot water is drawn as needed from **(D)**

FIGURE 6–3 Open and closed loop hot water systems.

freezing temperatures occur. Open loop systems are particularly advantageous in Florida, for example, where street pressure is low and freezing is rare.

In the closed loop system, potable water does not enter the collectors so the fluid used for heat transfer may contain antifreeze and corrosion inhibitors. A heat exchanger must be added to the system to transfer heat from the closed loop to the service loop and, in fact, some building codes require a double-wall heat exchange between the toxic closed loop and the potable service loop. This adds to the expense of a closed loop system but there are some advantages to be had. The pressure in a closed loop need not be any greater than that necessary to achieve circulation, and antifreeze can be used to provide protection. A closed loop system would be preferred in the mountains of Colorado where there are large variations in street water pressure and many freezing nights.

The design capacity of a liquid circulation pump is proportional to the collection area and will generally fall in the range of 10 to 30 pounds per hour per square foot of collector area, or 2 to 6 gallons per minute per hundred square feet of collector area. Operating pressure in the system will be reduced

111

FITTINGS

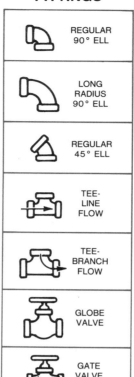

PRESSURE LOSS

FLOW (gpm)	COPPER OR STEEL TUBE		
	½"	¾"	1"
1	1.8	0.5	—
2	6.4	1.2	0.5
3	13.1	2.3	0.7
4	21.7	4.2	1.2
5	31.9	6.0	1.6
10	108.0	19.9	5.8
15	—	40.7	11.6

Units of ft-H_2O per 100 ft. of tube.

EQUIVALENT PIPE LENGTHS OF COMMON HYDRONIC FITTINGS[a]

	STEEL PIPE	COPPER TUBING
90° Elbow	25	25
45° Elbow	18	18
90° Elbow (long radius)	13	13
90° Elbow (welded)	13	13
Reducer coupling	10	10
Gate valve (open)	13	18
Globe valve (open)	300	425
Angle radiator valve	50	150
Radiator or convector	75	100
Boiler	75	100
Tee (typical)	100	100

Notes: a. Given in diameters of pipe.

FIGURE 6–4 Pressure loss in pipes and fittings.

through losses caused by system hardware; consideration of this potential loss can be made using the tabulated pressure loss for pipes and common fittings of various sizes contained in Figure 6–4.

THERMAL STORAGE

The storage of thermal energy is usually accomplished by taking advantage of the capacity of many materials to retain heat. Table D–6 lists the magnitude of this property, expressed as the more common specific heat, for some of the common materials either used or considered for use in solar energy systems.

Water is often used for thermal storage with an insulated, corrosion protected tank as its reservoir. The most common structural material for tanks is steel, protected either by rust inhibitors in the stored water or coatings on the steel surface, or both. Concrete tanks avoid the corrosion problem and are self-insulating, but they are more susceptible to leaks and are quite heavy. Alternative materials for tank construction include wood, fiberglass, and nonferrous metals, but none of these offers a significant advantage over steel or concrete. Depending upon the specific requirements at the site, compound materials (concrete mixed with plastic or fiberglass-lined wood tanks) may prove cost effective and suitable.

Thermal storage can also be accomplished by utilizing phase changes of a solid into a liquid or a liquid into a gas. The heat of fusion (latent heat) of a solid represents the heat required to melt it without raising its temperature. Similarly, the heat of vaporization of a liquid represents the heat required to vaporize the liquid without raising its temperature. Choosing the proper substance for this method of storage is predicated on service temperature; some materials do undergo phase changes at temperatures convenient for use. Rochelle salts, metallic Gallium, Glauber salts, paraffin wax, and fluorocarbons undergo phase changes appropriate for low and medium temperature regimes. Some eutectic alloys of metals, such as solder, and Wood's metal can also be used to store latent heat but the high temperatures at which these metals undergo phase changes are suitable only for concentrating solar cookers and not for most solar energy applications. Because of the increased costs and complexities of thermal storage by latent heat, these systems are not economical except where space limitations are severe. The latent heats of currently used or potential heat storage materials are also listed in Table D–6.

HEAT TRANSFER

Heat is usually transferred from the closed cycle working fluid to the storage or service fluid by means of a heat exchanger. Liquid to liquid heat exchange is generally accomplished by circulating one liquid through a coil of metal tubing within a tank or shell containing the other liquid with the rate of heat transfer approximately proportional to the temperature difference between the two liquids, the area of the total surface which separates the two liquids, and the overall heat transfer coefficient.

$$J = UA\Delta T_{lm} \tag{6–1}$$

where: J = rate of heat transfer in BTU/hr in the heat exchanger

U = overall heat transfer coefficient = 150/300 for forced circulation counter current flow (BTU/hr-ft²)

= 25/60 for natural circulation on one side (BTU/hr-ft²)

A = area of the heat exchanger surface (ft²)

$\Delta T_{lm} = \dfrac{t_{inlet} - t_{outlet}}{\ln \dfrac{t_{inlet}}{t_{outlet}}}$ = log mean temperature difference

t_{inlet} = difference between the water coming from storage and the fluid returning to the collectors (°F)

t_{outlet} = difference between the water returning to storage and the fluid coming from the collectors (°F)

Equation 6–1 is approximate because the rate of heat transfer is also influenced by the nature of the fluid flow (laminar flow results in less heat transfer than turbulent flow) and the degree of mixing of the fluid in the shell (which may be improved with proper baffling). For most applications, the direction of flow in the two liquids should be opposite to each other (in counterflow).

Heat exchangers are also made by constructing matrices of tubes that are interwoven and, sometimes, immersed in a heat conducting fluid. This technique can be applied to situations which require heat exchange between more than two fluids or in an exchange between fluids and air. Liquid to air heat exchangers, like the automobile radiator, are usually constructed of finned tubes; the variety of geometries utilized in these fin and tube exchangers is too diverse to permit brief exposition, so exchange capacity is best obtained from the manufacturer. Figure 6–5 illustrates a few of the available heat exchanger geometries.

The heat pipe and related products which utilize the vaporization, trans-

FIGURE 6–5 Typical heat exchange surfaces.

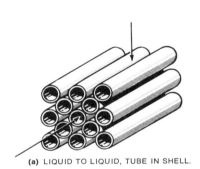

(a) LIQUID TO LIQUID, TUBE IN SHELL.

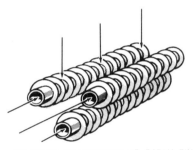

(b) LIQUID TO LIQUID OR AIR, RADIAL FIN.

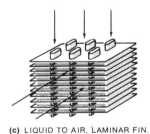

(c) LIQUID TO AIR, LAMINAR FIN.

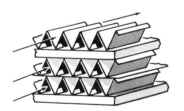

(d) LIQUID OR SOLID TO AIR, CORRUGATED FINS.

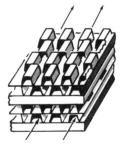

(e) LIQUID OR SOLID TO AIR, STAGGERED FINS.

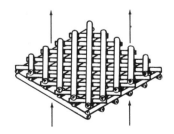

(f) LIQUID TO LIQUID OR AIR, TUBE MATRIX.

port and condensation of a heat transfer fluid in an enclosed, evacuated tube are efficient and capacious transfer devices and are used in applications ranging from cooling valves in high performance engines to home cooking utensils. Heat pipes have been proposed for many solar energy applications but cost factors have inhibited widespread acceptance. In addition, heat pipes are most effective in conveying heat from a small area to a large one, while most solar applications require just the reverse.

CONTROL TECHNIQUES

The automatic control of a solar thermal system is most often accomplished by controlling the fluid circulation pump in response to changes in system and service temperatures, to changes in the level of solar insolation, and to changes in ambient temperatures. The most common type of control system consists of a control box that turns the fluid circulation pump on whenever the solar collector temperature is greater than the thermal storage temperature and turns it off when it is less. Care must be taken to adjust the temperature limits and flow rate to prevent "cycling" where the action of turning on the pump causes the collector to cool, which subsequently causes the control to turn off the pump with the process continuing in a repetitive cycle. Control boxes should have high limit controls to prevent overheating and bypass controls to allow for manual operation. Some control systems provide a circuit for sensing freezing conditions and can be used to open a drain valve. Figure 6–6 illustrates some of the more common control circuits used in thermal solar collection systems.

THERMAL ENERGY CONVERSION

The conversion of thermal energy into higher forms of energy has been one of the principal objects of the science of thermodynamics ever since the discovery of a mechanical equivalent to heat energy in the seventeenth century. The laws of thermodynamics govern the terms upon which heat energy can be converted into other forms with two important rules emerging from the application of these laws. The first is that the processes which convert heat into other forms of energy are inherently less efficient than the processes which produce heat. The second is that the ability to convert heat into other forms of energy is dependent upon the temperature difference between two sources of heat with the corollary that heat energy cannot be extracted from one source of heat alone. The motive force for the conversion of heat to higher forms of energy is, therefore, the movement of heat from a hot object to a colder one.

The creation of mechanical energy or work from heat is accomplished in an engine. The expansion of a fluid upon vaporization is the basis of the Rankine cycle engine (more often known as the steam engine) and can be used to power a turbine, as well as a piston or any number of other ingenious mechanical arrangements. The efficiency of the Rankine or steam turbine cycle is usually less than that of the internal combustion engine, though efficiency does increase with an increase in steam temperature. These techniques may be useful in solar applications that involve concentrating collectors and large central station

FIGURE 6–6 Typical control circuits.

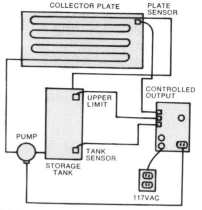

(a) PUMP CONTROL BY TEMPERATURE LIMITS

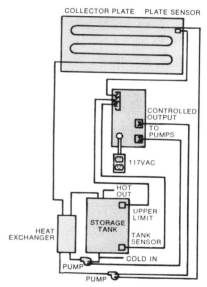

(b) DUAL PUMP CONTROL

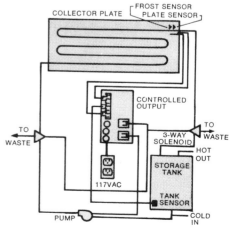

(c) CONTROL FOR FREEZE PREVENTION

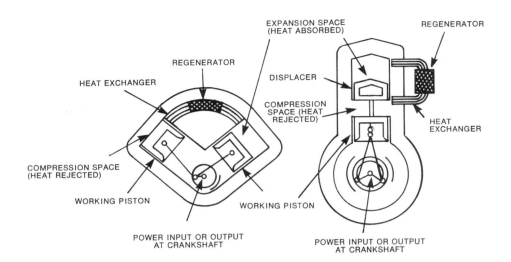

FIGURE 6–7 Stirling cycle engine. Two main types of Stirling engine: (1) left, double cylinder, two-piston; (2) right, single cylinder, piston-plus-displacer. Each has two variable-volume working spaces filled with the working fluid, one for expansion and one for compression of the gas. Spaces are at different temperatures—the extreme temperatures of the working cycle—and are connected by a duct, which holds the regenerator and heat exchangers.

facilities where high temperatures and pressures are possible, but they would seem to have little application in individual solar installations.

Heat supplied by flat plate solar collectors and other medium temperature sources may be converted into mechanical power through the use of a Stirling cycle engine (illustrated in Figure 6–7), which derives its power from the expansion of air as it is alternately heated and cooled.

Mechanical energy through motors powered by low temperature heat sources can make use of thermal effects such as the Minto wheel, bimetal expansion effects, and thermoelastic effects.

The conversion of thermal energy into electricity is usually accomplished by applying rotary mechanical power to an electrical generator shaft; electricity may also be produced directly from thermal energy through the use of thermoelectric devices which consist of junctions between dissimilar metals, or thermionic devices that emit electrons when heated in a vacuum. Small amounts of electricity can be produced by the effects of mechanical stress or thermally induced mechanical stress (such as that from the thermal expansion of metals or freezing water) acting on piezoelectric devices, or by the action of condensation in an electric field like that in lightning.

AIR CONDITIONING

As heat from hot water is ultimately conveyed to its surroundings, its energy can be utilized to produce a cooling effect. The principal methods of cooling are the mechanical vapor compression cycle, the absorption refrigeration cycle, and the vapor jet cycle, all of which are shown diagrammatically in Figure 6–8.

The mechanical compression cycle derives its input energy from mechanical work powering a compressor which heats a gas by compression. The heated gas is cooled to near ambient temperature and condensed in a heat

FIGURE 6–8 Cooling cycles.

(a) VAPOR COMPRESSION

(b) ABSORPTION

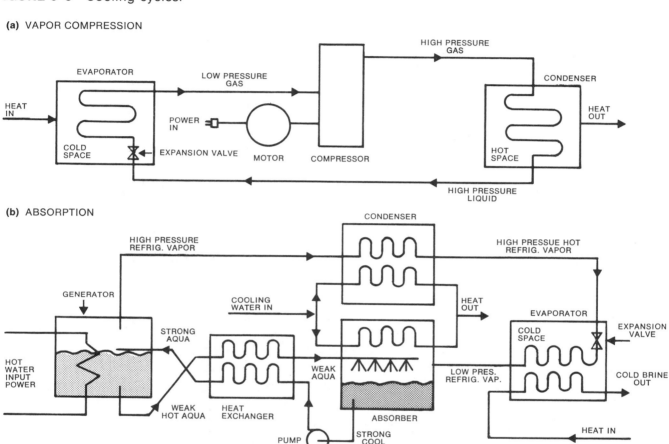

(c) VAPOR JET

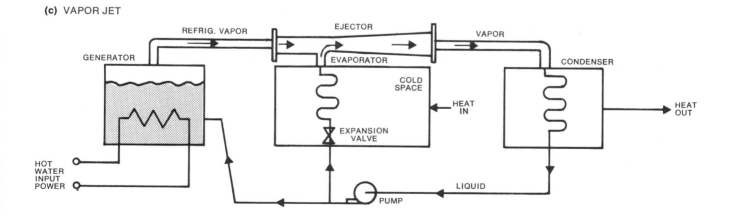

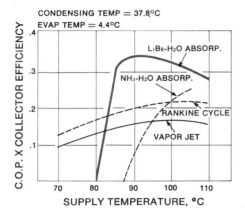

CONDENSING TEMP = 37.8°C
EVAP TEMP = 4.4°C

C.O.P. X COLLECTOR EFFICIENCY

LiBr-H2O ABSORP.

NH3-H2O ABSORP.

RANKINE CYCLE

VAPOR JET

SUPPLY TEMPERATURE, °C

FIGURE 6–9 Comparison of cooling cycle efficiencies.

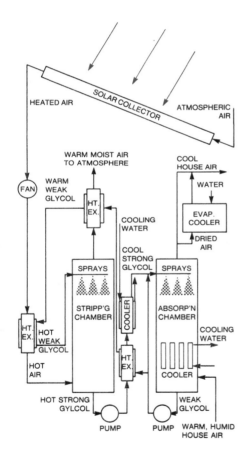

FIGURE 6–10 Cooling by solar-powered absorption dehumidification.

exchanger that is located outside the space to be cooled. Once cooled and condensed, the gas (or fluid) is routed through the application space where its expansion and vaporization absorbs heat from its surroundings to produce a cooling effect. In order for a solar energy system to supply power for a compression and expansion cycle, it would first have to convert its thermal energy into mechanical power. Because of this, the mechanical vapor compression cycle is suitable for use in solar systems with concentrating collectors which are capable of supplying temperatures high enough to provide the necessary mechanical power.

The absorption refrigeration cycle operates on the principle that the absorption capacity of gases in liquids is temperature and pressure dependent. For example, as a pressurized solution of ammonia in water is heated, it liberates ammonia gas at a high pressure which can then be cooled and condensed into a liquid. This refrigerant liquid ammonia is then allowed to expand into a low pressure region and vaporize, absorbing heat from the space to be cooled. The heated vapor is then reabsorbed into cool water, which is pressurized with a pump to complete the cycle. This absorption cycle is suitable for use in solar collecting systems which use high performance flat plate or concentrating collectors, but the system will usually require a cooling tower to convey heat to the environment. The lithium bromide absorption refrigeration cycle is similar to the ammonia cycle except that water, rather than ammonia, is the refrigerant that is absorbed into a solution with lithium bromide, not into water.

The vapor jet cycle makes use of the Venturi principle to produce a partial vacuum from an ejector which is driven by vapor passing from a boiler generator to a condenser. The partial vacuum is used to draw vapor from an evaporator which supplies the cooling effect. The vapor jet cycle can be used at lower temperatures and offers simplicity, but the cooling tower requirement is very large and the cycle's overall efficiency is poor.

A comparison of the thermal efficiencies of the various methods of cooling is shown in Figure 6–9. In this figure, the overall thermal efficiency is the product of the solar collector efficiency and the cooling cycle coefficient of performance (COP). The COP of a cooling cycle is the ratio of heat energy supplied to cooling effect produced and does not take into account mechanical or electrical energy inputs or cooling energy losses. Figures 6–10 and 6–11 depict two possible means of interfacing a cooling system with solar collectors.

HEAT PUMPS

A heat pump may be described as a mechanical vapor compression cycle operating between two reversible heat sources, i.e., in the summer it cools living space by transferring heat from building interiors to the outside, in the winter it performs the reverse by "air conditioning" (in a sense) the outside air and moving that extracted heat to living spaces.

Heat pumps have found wider acceptance in conserving energy with traditional fossil fuel systems, but they are also useful with solar energy systems. A solar energy system incorporating a heat pump can deliver usable energy over a much wider range of storage temperatures, and therefore conserve more energy, than a system without the heat pump assist. For example, a forced air heating system operating without a heat pump requires significantly higher minimum storage temperatures. With a heat pump, storage temperatures as low as 40°F may be boosted to temperatures high enough for heating use (90°F to 130°F). When storage temperatures drop below 40°F but ambient air temperature is above 40°F, the heat pump will use the outside air as its heat source and continue to contribute to increasing the storage temperature. The effect is equivalent to increasing the heat storage capacity and, because storage is usually at low temperatures, increasing the efficiency of the entire system. With ambient temperatures below 40°F, electric coil or oil fired burners switch on to provide heat directly to the house.

Cooling with solar-assisted heat pumps operates in a similar manner, but in the reverse. In addition, with some systems, the heat pump may be used at night to cool storage temperatures in preparation for the following day. There is no direct savings in energy use under this process since there is no overall reduction in energy consumption, but the electricity used to power the heat pump at night is consumed during low demand periods and, thus, can reduce the daytime peak demand load for utilities.

The decision whether or not to use a heat pump with a solar energy system is not a simple design choice. Analysis of the cost of the electrical energy needed to operate the heat pump during heating and/or cooling seasons versus its return in increased efficiency should be one basis for the decision.

DELIVERY TECHNIQUES

The delivery of solar thermal energy usually needs to be no different than the delivery used with other sources of energy. Hot water radiators, baseboard heating, and forced air ducts are all in common use. Since large wall and roof areas are often employed for collection purposes in solar heating and cooling, some delivery systems make use of these areas more directly; these systems are, however, generally more difficult to control. Conventional hot air delivery is illustrated in Figure 6–12, a typical hot water delivery system in Figure 6–13, and unconventional delivery techniques in Figure 6–14.

In air-conditioning applications, delivery is influenced by alternative methods of storage. Since the efficiency of the absorption cycle is very sensitive to the temperature of the solar water heater (see Figure 6–9), maximum use should be made of high temperature water directly from the collectors during those periods when it is available. At such times, the cooling apparatus is used to cool brine which is then stored. Heat storage is less efficient because it results in a lower supply temperature to the chiller than the collectors can provide directly.

For a combined heating and cooling system, two separate thermal storage tanks may be needed: one for hot water from the collectors, and another for cooled brine. If the solar system is remotely located from the serviced site, the absorption equipment should be as close as possible to the collectors.

The delivery of heated or cooled fluids in underground pipes or ducts requires thermal insulation which is protected from water, soil chemicals, and vermin. The techniques used require either special waterproof materials,

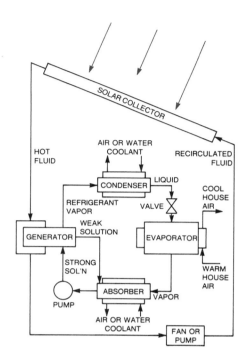

FIGURE 6–11 Cooling by solar-powered absorption refrigeration.

FIGURE 6–12 Conventional delivery of solar energy for hot water, space heating, and cooling with a circulating air system.

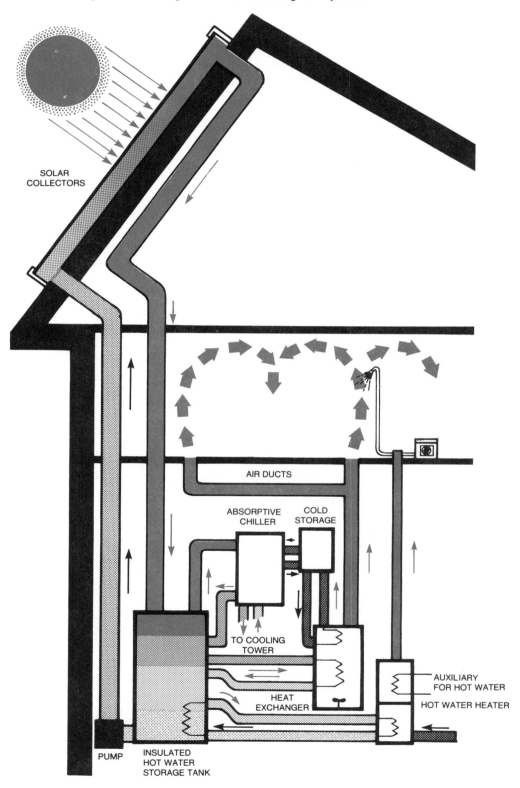

SOLAR COLLECTORS

AIR DUCTS

ABSORPTIVE CHILLER

COLD STORAGE

TO COOLING TOWER

AUXILIARY FOR HOT WATER

HEAT EXCHANGER

HOT WATER HEATER

PUMP

INSULATED HOT WATER STORAGE TANK

FIGURE 6–13 Conventional delivery of solar energy for hot water and space heating with circulating hot water.

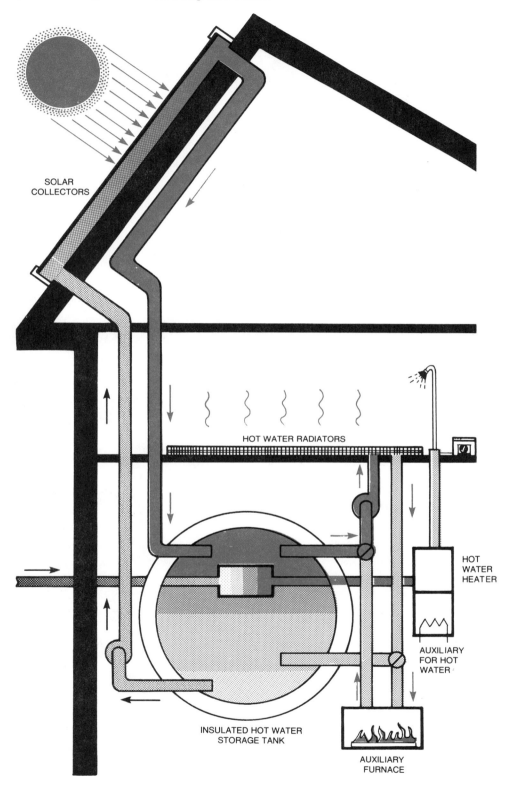

SOLAR COLLECTORS

HOT WATER RADIATORS

HOT WATER HEATER

AUXILIARY FOR HOT WATER

INSULATED HOT WATER STORAGE TANK

AUXILIARY FURNACE

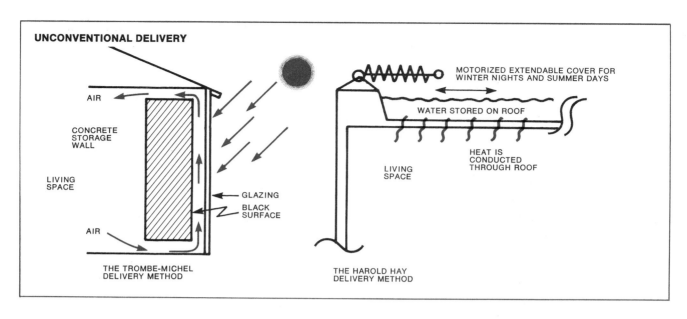

FIGURE 6–14 Unconventional delivery techniques.

pressurized conduit jackets, or pumped jacket drainage systems. Condensation of moisture can become a major problem with cooling ducts used in underground delivery unless watertight, low permeance, or vapor-barrier insulation methods are implemented.

AUXILIARY SYSTEMS

In the vast majority of applications, a solar energy system will be designed to provide a certain percentage of a given building's energy requirement, typically in the 50%–80% range, whether the system is intended for a minimum of providing domestic hot water or total service of space heating, cooling, and hot water. This is not to say that a solar energy system to supply 100% of the requirement could not be built; indeed, MIT successfully heated its first solar house with solar energy alone in 1939. But the costs of added system capacity, especially in increased collector area and storage capability, sufficient to counter periods of sunless days and cold weather can become unreasonable. In addition, the system would have to be designed to meet the most extreme conditions, an infrequent occurence, with a consequent loss in energy saved per square foot of collector area, since the system would be operating far below capacity at most times, and a lowering of return on the investment. As a result, an optimum solar energy system should be targeted toward that 50%–80% provision, with an integrated auxiliary system to fill in gaps during inclement weather or when thermal storage is exhausted.

The back-up system may be chosen from a wide range of alternatives. Oil or gas furnaces, electric resistance coils, and heat pumps have all been used; in retrofit solar installations, the existing system is simply incorporated through interfacing with the solar system. Efficiency of the entire system (solar and auxiliary) can be increased by integrating the two through shared components—pumps, piping, controls—to avoid redundancy in equipment and to reduce total cost.

One precaution should be observed in the interconnection of the solar energy and auxiliary systems. In most cases, the two systems should be connected in parallel, not in series. Back-up systems connected in series inadvertently heat the solar storage, causing an extra expense in energy and less efficiency through solar heat collection since the storage temperatures will be maintained at or near usable temperatures. In addition, maintenance or storage at higher temperatures through auxiliary heating will result in another inefficiency, since thermal losses from storage will increase.

SOME PRACTICAL CONSIDERATIONS

Because utilization of the solar resource as an alternative energy source is a relatively new field, the body of design expertise, installation experience, and operation knowledge is not widespread. This is not to say that such knowledge and expertise cannot be found; indeed, the sum of governmental, university, and commercial sources, and the solar literature, is comprehensive. Still, an individual planning to install a solar energy system, whether to heat the water in a residential swimming pool or to cool space air in an office building, cannot always rely upon finding local expertise. By comparison with an individual considering a new oil-fired heating plant or central electric air conditioning, where the variation of efficiency and reliability in available equipment is much smaller, and the telephone directory can supply a number of experienced contractors, solar energy utilization requires more effort and investigation.

Common sense, as with most pursuits, can make its contribution to a successful solar energy system, especially in equipment installation. For example, solar collection panels are often roof-mounted, which requires consideration of related concerns. Proper orientation to the sun is the most obvious, but provision for drainage of rain water under or around the panels to avoid standing water, mounting so that the underlying roof structure is not damaged, and integrity of the mounting so that the panels stay in place for the twenty or more years of their service life are some of the ancillary considerations. On the last point, it is worth considerable care to ensure that the panels are secure, since the effects of a heavy solar panel sliding from a tilted roof to the ground can be very serious.

Other elements of system design and installation may be accomplished by any number of alternative means and, consequently, are surrounded by a larger number of related considerations. System transfer fluids containing antifreeze solutions can have potentially hazardous effects, as has already been noted. The concern is real, so much so that many communities have instituted building standards requiring a double-walled heat exchanger between the collection fluid loop and the service loop to avoid contamination of one by the other. Similar guidelines and regulatory constraints—HUD standards on the federal level, and various local and state statutes—are in effect concerning solar energy installations.

SOLAR STANDARDS

One of the most frequently cited restrictions which limits the introduction of a new product or technology is the lack of complete standards by which to assess the value, quality and safety of the device or method. As a new technology, solar energy has, in addition to the usual institutional barriers, a unique set of complex component and system requirements which must be identified, qualified and, ultimately, certified. The problem lies in the basic fact that almost any solar device will work to one degree or another. But the answers to such questions as: How well does it work? How long will it last? and How safe will it be? require appropriate, and accepted, standards. Various groups have developed, or are developing, the standards required to answer these and other questions:

1. American Society of Heating, Refrigeration, and Air Conditioning Engineers (ASHRAE) has established standard testing procedures for solar collector thermal performance, thermal storage devices, solar domestic hot water systems, and solar swimming pool heaters.

2. National Bureau of Standards (NBS) has, in addition to providing valuable input to development of private consensus standards, been the primary developer of solar energy standards for the federal government. The Bureau's "Interim Performance Criteria" have been implemented in the HUD Minimum Property Standards, which include system sizing, collector descriptive information, and general installation techniques, among others.

3. American Society for Testing and Materials (ASTM) has formed, or proposed, a number of subcommittees to manage and administer the development of standards. The work of the subcommittee on materials performance, which is representative of other subcommittees, includes the development of testing standards for absorptive coatings, cover plates, fluids, reflective surfaces, insulation, metallic and nonmetallic containment, seals and gaskets.

4. Sheet Metal and Air Conditioning Contractor National Association (SMACNA) has developed an Institutional Standard which includes solar energy load calculation methods, equipment selection, and suggested installation practices.

5. Underwriters Laboratory (UL) operates labeling programs for solar related equipment which qualify the product's safety, but make no attempt to certify accuracy or performance characteristics. A safety standard for solar collectors, which will include fire and casualty risks associated with commercially built collectors, as well as safety measures for installation, maintenance and use, will be released by UL in the near future.

6. American Society of Mechanical Engineers (ASME), in its traditional role of standards development for pressure vessels, is addressing the safety of solar components and systems with respect to the high pressures and temperatures that can be developed in large arrays of flat plates, concentrating systems and power tower installations.

Certification programs, in which components or systems will be tested and their performance, maintenance and endurance characteristics qualified, are in progress in various quarters. These programs, once completed, will assist potential solar energy equipment purchasers in choosing appropriate equipment, as well as provide data on which governmental bodies may establish minimum standards in their jurisdictions. Some of the principal groups involved are:

1. International Association of Plumbing and Mechanical Officials (IAPMO) has published a "Uniform Solar Energy Code" which specifies material standards for pipe and fittings, and design and test pressure criteria for components. This code has been adopted by many local jurisdictions and two states as of February, 1979.

2. The City of Los Angeles has adopted the IAMPO as its basic code and added other qualifications to it. The city maintains a program of hydrostatic testing and construction inspection for collectors and other pressurized components. Certification under this program, which addresses health and safety issues and not thermal performance or system efficiency, is required for all installations within the city.

3. The State of California is instituting a certification program to rate collectors based on a "BTU per hour per square foot" measure as tested by independent laboratories. Mechanical testing for thermal shock-water spray, thermal shock-cold fill, and static pressure leakage will also be qualified. Though initially a voluntary program, proposals now under consideration would make certification a mandatory part of the California solar energy tax credit. A similar program, with required certification after January 1, 1980, is in practice in the State of Florida.

4. The Solar Energy Industries Association (SEIA) operates a parallel rating program to provide an all-day thermal rating for collectors tested under the ASHRAE standard procedures. The initial rating program will consider thermal performance only, with subsequent testing to include structural and mechanical integrity.

A laboratory certification program to qualify independent testing facilities has also been established.

These and other existing bodies of regulations and guidelines, codes and standards, undergo continual refinement or change, with new regulations added periodically. It may be reasonably assumed, therefore, that the numbers and complexity of applicable standards will increase, and that any listing of them soon becomes outdated. At the same time, and this cannot be emphasized too strongly, planning for any solar energy system must take cognizance of all federal, state and local regulations. Further, the tasks of equipment choice, system design, installation, operation, and maintenance should gain the advantages of appropriate qualification data and practical experience.

CHAPTER 7
DETERMINING COLLECTOR
AREA AND STORAGE
CAPACITY

As with most endeavors, there are methods of planning ranging from simple to complex which yield results, coherent to the method applied, which range from the merely approximate to the most accurate. In this chapter, the collector area and storage capacity of a sample solar energy system are determined through a procedure that is more accurate than the procedure used in Chapter 3, but which falls short of the accuracy possible through the use of large computers. Sources and magnitudes of potential error in this example procedure are noted so that the reader may make his own choice in practical situations of whether to follow Hopkins' dictum that "Genius is an infinite capacity for taking pains" or to proceed with system design with some degree of approximation. Having said that, it is well to note that solar energy systems are more costly to purchase and install than other energy systems and that, since collector area and storage capacity are the principal design features and those most difficult to alter by afterthought, the greatest need for accuracy exists in the determination of these two components.

The capacity for heating and cooling of any system is specified by the number of BTUs per hour or per day required to achieve the desired effect. BTU requirements can be estimated by calculation or by measurement for existing buildings; for those structures still in the planning stage, a calculation must suffice. Once the BTU requirements are known, the solar system design can begin. Collector area is determined by site geography and geometry, solar insolation statistics, collector efficiency, climatic variables, and service temperature requirements. As a general rule, collector area for space heating will fall in the range of one-quarter to one-half of the floor area; hot water service and hot water storage requirements fall in the range of one to two gallons for each square foot of collector area; hot air service requires one-third to two-thirds cubic feet of rocks for each square foot of collector area.

An important characteristic that distinguishes a solar energy system from other energy supply systems is the comparative lack of flexibility in supplying variable loads over extended time periods. The thermal storage of a solar energy system is limited in both capacity and time; it cannot supply more than its peak collection capacity for a time longer than its thermal storage will allow—rarely more than one week. The collection of solar energy for storage, which is accomplished during periods of slack demand, is similarly limited to that time period when solar insolation is usable. By contrast, conventional fuel systems are not limited in this manner, since one may draw heavily upon them, or not at all, for an extended period. By the same token, fuel storage in a conventional system can be of long duration, neither diminishing with time nor dissipating when it is not being used. The implication within this lack of flexibility in solar systems is the possibility of an occasional, infrequent, shortfall of energy supply. It is a matter for consideration, hence the requirement for back-up conventional systems, but solar shortfalls are no cause for undue concern. Conventional fuel systems have experienced similar shortages in the past, will continue to do so more often in the future, with ever-increasing cost of new supply, until the last barrel of oil or cubic foot of natural gas is pumped from the ground. That those homes and buildings with solar energy systems will see that inevitable day arrive without concern for the next delivery to these systems is as dependable as the morning's sunrise and the costs for the sun's input can never rise above the single zero.

Another matter for consideration is the question of overdesign. It is common practice in the planning of conventional fuel energy systems to provide a significant level of overdesign, sometimes as much as 100% more than peak needs require. This provides a healthy margin of error in the computation of energy needs which, since the initial costs of a conventional system are lower, is affordable. In the planning of a solar energy system, overdesign is very costly because it results in poor utilization of available solar insolation. Depending upon the control strategy used in the system, overdesign in collector area may result in increased stagnation time, or the necessity to incorporate heat wasters, or excess temperatures, in addition to higher costs for collectors, support structures, and interconnection headers. Overdesign of thermal storage capacity can lead to insufficient temperatures, as well as excessive weight, space utilization waste, and cost. The question is the eternal problem of drawing the dividing line between too much and too little. Striking the middle ground by accurately fitting collector area and storage requirements to the average demand is the best recommendation. Back-up systems using conventional fuels or electric power should be used for overdesign capacity for those periods when solar insolation becomes unusually weak, or demand becomes unusually large, for an extended period of time.

CALCULATING BTU REQUIREMENTS

The most comprehensive procedure for calculating the BTU requirements for space heating and air conditioning is found in the ASHRAE *Handbook of Fundamentals*. This volume of more than 650 pages covers most of the fine details necessary to make a highly accurate computation of BTU requirements. For the purposes of this handbook some of the more important features of the detailed procedure are abstracted and simplified to yield an approximate calculation for BTU requirement. It should be noted that BTU requirements are influenced by lifestyle and variations among individuals' behavior account for substantial differences in energy consumption. The occupant of a house who sets the temperature at 80°F and leaves the windows and doors open in the winter while consuming large amounts of hot water by frequently bathing and washing dishes and clothes may consume more than twice the energy of one who wears sweaters in 65°F temperature, keeps the doors and windows sealed shut, and washes only once or twice a week. The examples are extreme but correspond to greatly different energy needs. BTU requirements here are calculated on the basis of average or statistical norms of behavior, and each individual must make his own adjustments to accommodate those personal factors which consume or conserve energy.

HOT WATER

The temperature of hot water should be in the range of 110°F to 150°F, with most hot water heaters set to begin heating when the water temperature is 130°F or below and to shut off when 150°F is reached. The individual user then mixes hot water with cold to gain the desired temperature. Minimum hot water storage tank capacities and BTU requirements, assuming the inlet water is 60°F and that the storage tank is not quite emptied twice each day, are specified by the U.S. Department of Housing and Urban Development (HUD) and the Federal Housing Authority (FHA) for one and two-family living units as shown in Table 7–1.

SPACE HEATING AND COOLING

The BTU requirements for heating and cooling are calculated by determining the heat loss of a structure that must be made up by heating or cooling. Heat loss is the product of the temperature difference between interior and exterior air temperatures, the area of the surface, and its U-factor, with loss considered outward for heating and inward for cooling. The U-factor is the thermal conductivity (k-factor) in BTU-in/hr-ft²-F° divided by the thickness (in inches) of the insulating material.

$$U = \frac{k}{t} = \frac{1}{R} \text{ BTU/hr-ft}^2 \text{ °F} \tag{7–1}$$

where: U is the U-factor of a wall (BTU/hr-ft² °F)

      k is the thermal conductivity of a material in BTU · in/hr-ft² °F

      t is the thickness of a wall in inches

      R is the thermal resistance of the wall.

TABLE 7–1   HUD-FHA Hot Water Requirements

	1–1½ Baths			2–2½ Baths				3–3½ Baths			
Number of bedrooms	1	2	3	2	3	4	5	3	4	5	6
Minimum storage	20	30	30	30	40	40	50	40	50	50	50
BTUs required per day (thousands)	27	36	36	36	36	38	47	38	38	47	50

The heat loss, Q, of a structure can be estimated by adding the heat losses for all exterior surfaces of the structure.

$$Q = Q_{\text{walls}} + Q_{\text{ceiling}} + Q_{\text{floor}} + Q_{\text{windows}} + Q_{\text{doors}} \text{ (BTU/hr)} \qquad \textbf{(7-2)}$$

$$Q_{\text{wall}} = A_{\text{wall}} \cdot U_{\text{wall}} \cdot (T_{\text{in}} - T_{\text{out}})$$
$$Q_{\text{ceiling}} = A_{\text{ceiling}} \cdot U_{\text{ceiling}} \cdot (T_{\text{in}} - T_{\text{attic}})$$
$$Q_{\text{floor}} = A_{\text{floor}} \cdot U_{\text{floor}} \cdot (T_{\text{in}} - T_{\text{basement}}) \qquad \textbf{(7-3)}$$
$$Q_{\text{windows}} = A_{\text{windows}} \cdot U_{\text{window}} (T_{\text{in}} - T_{\text{out}})$$
$$Q_{\text{doors}} = A_{\text{doors}} \cdot U_{\text{door}} (T_{\text{in}} - T_{\text{out}})$$

where the Qs are heat losses in BTUs per hour, the As are areas in square feet, the Us are U-factors in BTU/hr-ft^2-°F, and the Ts are temperatures in °F.

U-factors for various building materials are sometimes provided by their manufacturers or dealers, but the R-factor is more common. Converting R-factors to U-factors is easily done, using Equation 7–1.

Two alternative practices in construction require an additional first step in the calculation. If walls are made of layers of different materials, the Rs are added to obtain an R for the combination. If different walls are made of materials with different Rs, then the heat loss must be calculated for each area separately and then added.

The heat loss calculated by Equation 7–1 assumes that windows and doors are airtight and always closed. Of course, additional heat loss occurs by infiltration through open doors, windows, and cracks at a rate that depends upon wind speed and the size of the cracks. The actual heat loss from a real door is often better represented by its perimeter than by its area, since more heat is usually lost through cracks than through the area of the door.

A reasonable approximation to the total heat loss of a residential structure can be obtained by ignoring the inner wall and ceiling surface material and using only the U-factors of the insulation and exterior building materials in the calculation for heat loss. The result thus computed will be larger by nearly the amount that ignoring the effects of infiltration will make it smaller. For more detailed and accurate methods of calculating heat loss, the references on this subject are recommended (see Bibliography).

Heat loss calculated from Equation 7–2 provides both a peak BTU requirement (when the lowest outdoor temperature is used) and an average BTU requirement (when the average outdoor temperature is used). The monthly BTU average requirement can be obtained by substituting the number of degree days from Table 3–3 in place of the temperature difference factors in Equation 7–3 and then multiplying the result by 24 hours per day. That is:

$$Q_{\text{month}} = Q \times \frac{DD}{T_{\text{in}} - T_{\text{out}}} \times 24 \text{ BTU/month} \qquad \textbf{(7-4)}$$

where: Q_{month} is the heat loss per month

DD is the number of degree days in the month from Table 3–3

T_{in} is the desired indoor temperature

T_{out} is the average ambient temperature from Table 3–5.

MEASURING BTU REQUIREMENTS

In an existing structure, the BTU requirement can be approximately measured by multiplying the electric power or fossil fuel consumed in supplying energy to the structure for a day or for an entire heating season by the appropriate conversion factor. The factor for electric heat is $1\,Kw = 3413$ BTU/hr while that for oil heat (assuming that #2 heating oil is used) is 140,000 BTUs for each gallon burned. In a typical oil heating system, however, as much as 35% of the heating value is lost through escape in the flue, which carries combustion products outside the structure. As a round number, it is more accurate to regard one gallon of heating oil as equivalent to the heating effect of 100,000 BTUs. Each cubic foot of natural gas contributes about 1,000 BTUs with more efficient combustion than oil, so that about 800 BTUs are effectively provided. The recoverable heat energy from coal is about 10,000 BTUs per pound or approximately 20 million BTUs per ton. Table 7–2 summarizes these approximate figures.

PROCEDURES FOR ESTIMATING AREA AND CAPACITY

STEP 1

Locate the solar site on the maps in Figures 3–1 through 3–12 and select the nearest city among those listed in Tables 3–1, 3–3, and 3–5. Enter the values, which may be rounded to two significant figures, from these figures and tables on a monthly worksheet, as shown in Table 7–3. (The example used is based on a site near Philadelphia.)

If the site is not within 75 miles of one of the cities in the tables, select values from two other cities in opposite directions from the site and interpolate to estimate the value at the site. One should use common sense in situations where a site location is on a windy hilltop or in a misty valley, or obtain detailed climatological data when available, to make the necessary adjustments to conform to experience.

TABLE 7–2 Approximate BTU Equivalents among Energy Sources

Source	Unit	Approximate Heating Values in BTUs	
		Theoretical	Practical*
Solar power	One square foot of bright sun for one hour at solar noon	320	220
Electric power	One kilowatt hour	3,413	3,413
#2 Fuel oil	One gallon	143,000	100,000
Natural gas	One cubic foot	1,000	800
Coal	One pound	1,300	1,000

* After system losses are considered in heating applications.

Examine the site geometry in relation to the sun. For many applications it is only necessary to select a southward facing roof or wall with exposure to the sun from 9:00 AM to 3:00 PM in the winter. Southern exposure is strongly recommended for the obvious reason, but orientation of the collector is not so critical that a minor change in direction from due south, necessitated by the realities of the site or installation error, will seriously affect the system's performance. To illustrate this point, the example used in this chapter is based upon collectors oriented at 15° away from due south toward the east, and at a 5° steeper tilt than the latitude at the Philadelphia site, even though the general rule is to orient the collector toward due south with a tilt angle equal to the site's latitude. Furthermore, the realities of any site may require a change in collector orientation; the example departs from the general rule for orientation to accommodate the influence of Philadelphia's typical weather, i.e., sun in the morning with afternoon clouds. In brief: the "optimal" orientation of collectors (facing due south at a tilt angle equal to site latitude) may not always be the optimal orientation at the location under study. The following general observations can be of assistance in determining the most appropriate collector orientation.

Hot Water Service. Hot water is required throughout the year and at various times during the day. Tilt the collector at approximately the angle of latitude and perhaps slightly easterly from due south to catch the morning sun. If some seasons require awakening before dawn, tilt the collector westerly to catch the late afternoon sun and insulate the storage tank well.

TABLE 7–3 Site Climate Worksheet

City: Philadelphia
Latitude: 40° N
Longitude: 75° W
Elevation: 200 ft

	Jan.	Feb.	Mar.	Apr.	May	June	July	Aug.	Sept.	Oct.	Nov.	Dec.	Average All Year
Solar insolation (Ly/Dy)	150	225	300	400	475	500	525	450	350	250	150	125	325
×3.69 (BTU/hr-ft²) (from Figs. 3-1–3-12)	550	830	1,110	1,480	1,750	1,850	1,940	1,660	1,290	920	550	460	1,200
Average ambient temperature (°F) (from Table 3–5)	32	34	42	53	63	72	77	75	68	57	46	35	55
Heating degree days	1,000	870	720	370	120	0	0	0	40	250	560	920	404
Average % sunshine (from Table 3–1)	50	53	57	56	58	63	62	63	60	59	52	50	57
Average wind speed (mph) (from Table 3–10)	10	11	12	11	10	9	8	8	8	9	10	10	10

NOTE: 1 Ly/Dy = 3.69 BTU/ft²/day.

Space Heating Service. Since heat is needed primarily in the winter, and most abundantly in the morning, tilt the collector toward the peak elevation of the sun in late January, latitude $+15°$ (again, more or less). It may perform better if facing somewhat westerly from due south because solar collectors become more efficient as the outdoor temperature rises in the afternoon. This possibility for increased efficiency is obviously negated if late afternoon cloudiness is common at the site, an example of the manner in which implementation theory must be combined with consideration of practical realities before decisions are made.

Air Conditioning Service. Air conditioning is utilized primarily in the summer and most abundantly in the afternoon and evening. Tilt the collector at the peak elevation of the sun in late July, latitude $-15°$ (more or less). Performance may be improved if the collector faces slightly easterly from due south, especially in locations where mornings are usually clear, but afternoon cloudiness is frequent.

The optimum tilt orientation of a collector is neither solely nor critically determined by the tilt necessary to face the sun at its peak elevation; experience and the effects of related factors can alter the theoretical position. As an

---

TABLE 7–4 Site Geometry Worksheet

City: Philadelphia

Latitude: 40°

Longitude: 75°

Tilt Angles:
From horizontal
$\Sigma = 45°$
From direction of due south
$\phi = +15°$

A = Elevation angle of the sun (Equation 2–9)
Z = Azimuth angle of the sun (Equation 2–11)
θ = Incident angle (Equation 2–12)

On the 21st Day of

Solar Time	Dec.			Jan. Nov.			Feb. Oct.			Mar. Sept.			Apr. Aug.			May July			June		
	A	Z	Cos θ	A	Z	Cos θ	A	Z	Cos θ	A	Z	Cos θ	A	Z	Cos θ	A	Z	Cos θ	A	Z	Cos θ
4:00																					
5:00																2	115	0	4	117	0
6:00													8	99	.17	13	106	.15	15	108	.14
7:00							4	71	.44	11	80	.43	19	90	.41	24	97	.38	26	100	.37
8:00	5	53	.62	8	55	.63	14	61	.65	23	70	.65	30	79	.62	36	87	.59	37	91	.57
9:00	14	42	.78	17	44	.80	24	49	.82	33	57	.82	41	67	.79	47	76	.75	49	80	.73
10:00	21	29	.89	24	31	.91	31	35	.93	42	42	.94	51	51	.91	58	61	.86	60	66	.84
11:00	25	15	.94	78	16	.96	36	18	.99	48	23	.99	59	29	.96	66	37	.91	69	42	.89
12:00	27	0	.93	30	0	.94	38	0	.97	50	0	.98	62	0	.95	70	0	.90	73	0	.87
1:00	25	−15	.85	28	−16	.87	36	−18	.89	48	−23	.90	59	−29	.87	66	−37	.82	69	−42	.80
2:00	21	−29	.72	24	−31	.73	31	−35	.76	42	−42	.76	51	−51	.73	58	−61	.69	60	−66	.67
3:00	14	−42	.54	17	−44	.55	24	−49	.57	33	−57	.56	41	−67	.54	47	−76	.51	49	−80	.49
4:00	5	−53	.33	8	−55	.34	14	−61	.34	23	−70	.33	30	−79	.31	36	−87	.29	37	−91	.28
5:00							4	−71	.09	11	−80	.08	19	−90	.06	24	−97	.05	26	−100	.04
6:00													8	−99	0	13	−106	0	15	−108	0
7:00																2	−115	0	4	−117	0
8:00																					

example, the convective losses from flat plate collectors tend to decrease with larger tilt angles, so greater efficiency might be pursued with greater tilt. A steep tilt angle, however, incurs a more severe wind load. Also, a smaller tilt angle in the summer will allow the sun to spend more time in front of the collector rather than behind. The final decision must, again, rest upon a combination of theoretical calculations and practical sense.

In any case, the collector angle is specified by the azimuth angle, ϕ, of the normal which is the amount by which the collector tilt is directed away from due south. Due south exposure, then, is represented as $\phi = 0$. The elevation angle, ψ, of the collector normal is the 90° complement to the tilt angle, Σ, so $\Sigma = 90° - \psi$. Figure 2–6 contains a summary of the detailed geometric relationships. Good year-round exposure can be achieved when $\Sigma = L$, the latitude of the site and $\phi = 0$ representing exposure due south.

Once the collector normal angles, ϕ and ψ, are selected (either by design or fortuitous choice), one may proceed. A Site Geometry Worksheet like that in Table 7–4 can be used to describe the angles and the shading which form the relationship between the collector and the sun's apparent motion for each hour of the day and each month of the year. This example uses $\phi = 15°$ (facing east by 15° from due south) and $\psi = 45°$ (5° steeper tilt than the latitude at Philadelphia) with the entries in the worksheet taken from Equations 2–9, 2–11, and 2–12. A programmable calculator will facilitate the calculations. The blank areas on the worksheet represent those periods before sunrise and after sunset. Zeros are entered in the worksheet whenever $\cos \theta$ is negative (corresponding to those times when the sun is behind the collector) and for those hours or months when the collector is shaded by buildings, hills, trees, or other objects.

STEP 3

Determine the total solar insolation—direct, diffuse, and reflected components—incident upon the collector. All these components incident on a horizontal surface averaged over sunny and cloudy days are included in the daily solar insolation figures on the Site Climate Worksheet from Step 1, but as a total without distinction among them. The air mass, m, used in the Total Insolation Worksheet is derived by Equation 2–15, using the values of solar elevation, A, shown in the Site Geometry Worksheet. The direct normal solar insolation, I_n, is calculated from Equation 2–14 and recorded in the Total Insolation Worksheets (Tables 7–5 through 7–8). Next, the beam insolation, I_b, on the tilted collector surface is calculated from Equation 2–13 and the values of $\cos \theta$ gained from the Site Geometry Worksheet. Then, if the collector characteristics are known, the incident angle modifier is provided by multiplying I_b by $K_{\tau\alpha}$ from Equation 4–6. Finally, the total insolation, I_T, is determined from Equation 3–1.

The total insolation obtained on the Total Insolation Worksheet represents the insolation that is obtained on a clear, sunny day. It is generally less than the solar insolation that occurs on a sunny, though partially cloudy, day because it ignores the contribution of diffuse radiation from clouds, which can be considerable. Equation 3–6 can be used to obtain a more accurate value for diffuse radiation when cloud cover data is available but, if not, there is no reason for

TABLE 7–5 Total Insolation Worksheet (Winter)

Solar Time	December 21					January 21					February 21				
	m	I_n	I_b	$I_b \times K_{\tau\alpha}K_s$	I_T	m	I_n	I_b	$I_b \times K_{\tau\alpha}K_s$	I_T	m	I_n	I_b	$I^b \times K_{\tau\alpha}K_s$	I_T
6:00															
7:00											14.2	48	21	18	22
8:00	11.4	79	49	45	52	7.1	145	91	85	97	4.1	212	138	128	147
9:00	4.1	220	171	165	184	3.4	244	195	189	210	2.4	270	222	216	239
10:00	2.8	265	236	232	255	2.4	279	254	251	275	1.9	292	271	269	294
11:00	2.4	281	265	262	287	2.1	292	281	279	304	1.7	302	299	299	325
12:00	2.2	288	268	265	290	2.0	298	280	277	303	1.6	305	296	295	322
1:00	2.4	281	239	234	258	2.1	292	254	249	275	1.7	302	269	265	291
2:00	2.8	265	191	182	204	2.4	279	209	194	218	1.9	292	222	213	238
3:00	4.1	220	119	106	125	3.4	244	139	121	142	2.9	270	154	140	163
4:00	11.4	79	26	19	26	7.1	145	99	37	50	4.1	212	72	55	73
5:00											14.2	48	4	0	3
6:00															
Total Daily Insolation:				*1681*					*1874*					*2117*	

TABLE 7–6 Total Insolation Worksheet (Spring)

Solar Time	March 21					April 21					May 21				
	m	I_n	I_b	$I_b \times K_{\tau\alpha}K_s$	I_T	m	I_n	I_b	$I_b \times K_{\tau\alpha}K_s$	I_T	m	I_n	I_b	$I_b \times K_{\tau\alpha}K_s$	I_T
6:00						7.1	103	17	7	19	4.4	151	23	7	28
7:00	5.2	163	70	59	74	3.1	212	87	71	97	2.4	221	84	67	98
8:00	2.5	250	153	152	176	2.0	256	159	147	177	1.7	255	151	138	173
9:00	1.8	281	230	224	251	1.5	279	220	213	246	1.4	272	204	196	234
10:00	1.5	297	279	277	306	1.3	291	264	261	296	1.2	282	243	238	277
11:00	1.3	304	301	301	330	1.2	297	285	283	319	1.1	287	261	258	298
12:00	1.3	306	300	299	329	1.1	299	284	282	317	1.1	289	260	256	297
1:00	1.3	304	274	270	299	1.2	297	258	253	289	1.1	287	235	229	269
2:00	1.5	297	226	217	246	1.3	291	212	202	237	1.2	282	195	184	223
3:00	1.8	281	157	142	169	1.5	279	150	134	168	1.4	272	139	122	160
4:00	2.5	250	83	62	86	2.0	256	79	57	88	1.7	255	74	51	87
5:00	5.2	163	13	0	16	3.1	212	13	0	25	2.4	221	11	0	31
6:00						7.1	103	0	0	12	4.4	151	0	0	21
Total Daily Insolation:				*2282*					*2290*					*2196*	

TABLE 7–7 Total Insolation Worksheet (Summer)

Solar Time		June 21					July 21					August 21			
	m	I_n	I_b	$I_b \times K_{\tau\alpha}K_s$	I_T	m	I_n	I_b	$I_b \times K_{\tau\alpha}K_s$	I_T	m	I_n	I_b	$I_b \times K_{\tau\alpha}K_s$	I_T
6:00	3.8	159	22	5	29	4.4	141	21	6	28	7.1	86	15	6	18
7:00	2.3	219	81	64	97	2.4	211	80	64	96	3.1	193	79	65	92
8:00	1.7	249	142	128	166	1.7	246	145	132	170	2.0	238	148	136	170
9:00	1.3	266	194	185	225	1.4	263	197	189	229	1.5	262	207	200	237
10:00	1.1	275	231	226	267	1.2	273	235	230	272	1.3	274	249	246	285
11:00	1.1	280	249	245	288	1.1	278	253	250	292	1.2	281	269	268	307
12:00	1.0	282	245	240	283	1.1	280	252	248	291	1.1	282	268	267	306
1:00	1.1	280	224	217	259	1.1	278	228	222	264	1.2	281	244	240	279
2:00	1.1	275	185	173	215	1.2	273	188	178	220	1.3	274	200	191	229
3:00	1.3	266	130	113	154	1.4	263	134	118	158	1.5	262	141	126	163
4:00	1.7	249	70	47	85	1.7	246	71	49	87	2.0	238	74	53	87
5:00	2.3	219	9	0	33	2.4	211	11	0	32	3.1	193	12	0	27
6:00	3.8	159	0	0	24	4.4	141	0	0	21	7.1	86	0	0	12
Total Daily Insolation:					2125					2160					2212

TABLE 7–8 Total Insolation Worksheet (Autumn)

Solar Time		September 21					October 21					November 21			
	m	I_n	I_b	$I_b \times K_{\tau\alpha}K_s$	I_T	m	I_n	I_b	$I_b \times K_{\tau\alpha}K_s$	I_T	m	I_n	I_b	$I_b \times K_{\tau\alpha}K_s$	I_T
6:00															
7:00	5.2	140	60	50	66	14.2	35	15	13	16					
8:00	2.5	228	148	138	164	4.1	189	123	115	134	7.1	131	83	77	89
9:00	1.8	260	213	207	237	2.4	250	205	199	224	3.4	231	185	179	200
10:00	1.5	276	260	258	290	1.9	272	253	251	277	2.4	267	243	240	264
11:00	1.3	284	281	281	313	1.7	283	280	280	308	2.1	280	269	268	293
12:00	1.3	286	280	280	313	1.6	287	278	277	305	2.0	286	269	266	292
1:00	1.3	284	256	252	285	1.7	283	252	248	276	2.1	280	244	239	265
2:00	1.5	276	210	202	234	1.9	272	207	199	225	2.4	267	195	186	210
3:00	1.8	260	145	131	161	2.4	250	142	129	154	3.4	231	127	114	135
4:00	2.5	228	75	56	82	4.1	189	64	49	67	7.1	131	45	34	46
5:00	5.2	140	11	0	16	14.2	35	3	0	3					
6:00															
Total Daily Insolation:					2161					2039					1794

concern since the performance predicted by the procedure applied here will be conservative.

In the example shown in the worksheets, a value of $K_s = 1$ was used. This is equivalent to neglecting the effect of shadows from the walls of the collector assembly on the absorber plate. Note that the values of direct normal insolation, I_n, may not always agree with those tabulated by the ASHRAE method, despite the fact that similar methods are used in calculation. The differences, which are small, are due to the use of differing values for the solar constant (429 BTU/hr-ft² versus 428 BTU/hr-ft² in ASHRAE) and the use of formulas given by Equations 2–16 and 2–17 for I_r and B versus the use of tabular values in ASHRAE. The continuous equations avoid the step changes in values of I_n as one crosses from one month to another and when considering its day to day changes.

STEP 4

Determine the average daily solar energy available for each month of the year. This is a simple calculation which utilizes the data already collected. The Total Insolation Worksheet shows the total hourly amount of solar energy that can be collected for each month of the year. It reflects the realities of the variations of site location and collector tilt angles and provides information on the diffuse and reflected energy portions of the energy incident to the collector surface. The daily amount of solar energy is obtained by adding the hourly contributions on the Total Insolation Worksheet; the monthly figures are obtained by multiplying the daily amount by the number of days in each month. This is shown in the Insolation Summary Worksheet, Table 7–9. The monthly figures are then multiplied by the average percent sunshine figures from the Site Climate Worksheet developed in Step 1. The result is the amount of solar

TABLE 7–9 Insolation Summary Worksheet

City: Philadelphia
Latitude: 40°
Longitude: 75°
Elevation: 200 ft

Collector tilt angles: From horizontal, $\Sigma = 45°$
From due South, $\phi = +15°$
Reflection coefficient: $C_r = 0.25$ (Grass)

	Jan.	Feb.	Mar.	Apr.	May	June	July	Aug.	Sept.	Oct.	Nov.	Dec.
# of days	31	28	31	30	31	30	31	31	30	31	30	31
Total daily insolation BTU/day-ft²	1874	2117	2282	2290	2196	2125	2160	2212	2161	2039	1794	1681
Total monthly insolation 1000 BTU/mo-ft²	58.1	59.3	70.7	68.7	68.1	63.8	67.0	68.6	67.0	63.2	53.8	52.1
Average % sunshine (% ÷ 100)	.50	.53	.57	.56	.58	.63	.62	.63	.60	.59	.52	.50
Average available solar energy 1000 BTU/mo-ft² —on tilted surface	29.1	31.4	40.3	38.5	39.5	40.2	41.5	43.2	40.2	37.3	28.0	26.1
—on horizontal surface	17.1	23.2	34.4	44.4	54.3	55.5	60.1	51.5	38.7	28.5	16.5	14.3

energy that is available under those average conditions of sunny and cloudy days at the site.

Note that in this example the variation of average daily solar insolation on the tilted surface is less than ±10% of 40,000 BTU/mo-ft² for the months from March through October. At the same time, the horizontal daily insolation varies over a wide range. This indicates that the tilted surface is suited for applications that require a uniform consumption of heat during that period (as will be necessary in the supply of domestic hot water) and that during the winter months, despite the beneficial tilt angle, supplementary sources of energy may be required because of the relative reduction in available solar energy and the cooler ambient temperatures in which the collector must operate.

The value of tilting the collector can be seen by comparing the horizontal insolation values from the Site Climate Worksheet of Step 1 (in Langleys per day) to those in the Insolation Summary Worksheet. This is done by multiplying the horizontal insolation values by 3.69 to convert to BTU/ft²-day and then multiplying by the number of days in the month. For the example, this comparison shows that more energy is incident on the tilted surface in the winter and less in the summer than would be incident on a horizontal surface. The amount of improvement will depend on the tilt angles and the month of the year.

STEP 5

The performance of a flat plate collector must be evaluated. The thermal performance of solar collectors is measured and reported as efficiency, i.e., the proportion of incident solar energy which is captured and delivered to useful application.

Efficiency as a description of performance may be calculated through NBSIR 74-635 (based on aperture area and average fluid temperature) or ASHRAE 93-77 (based on gross collector area and inlet fluid temperature); regardless of the test method, the same number of collectors will be required for a given application.

But, as was shown in Chapters 4 and 5, efficiency is not a single number because it is dependent on operating conditions which vary constantly. Ambient temperature changes throughout the year, rises and falls throughout each day. Solar insolation characteristics, both intensity and incident angle, also differ with each season and, within each day, strike the collector at a succession of decreasing, then increasing, angular values around the noon hour. The temperature in the system's thermal storage facility, another influence on collector performance, is rarely constant.

To simplify the calculations, therefore, storage temperature will be assumed constant, and the complexity of the collector's off-angle performance will be replaced by the assumption that the collector is operating only on the noon curve. The performance estimate thus calculated will put rough bounds on system size for the example site, and allow comparison of different collectors on an equal basis. Obviously, where an actual application is being sized, good practice demands a thorough analysis of actual conditions with no generalized assumptions. The NBSIR 74-635 method of calculating efficiency is used here.

In the example, the AMETEK high performance collector efficiency curve, Figure 7-1, is used to determine the average collector efficiency. The

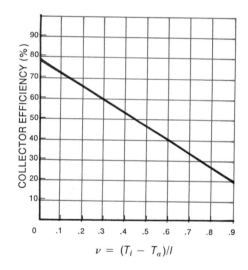

FIGURE 7-1 Performance Data. Collector panel of copper coated with a selective surface ($\alpha = 0.98$, $\epsilon = 0.30$). Double glazed with glass having a transmission, $T = 0.90$

Theoretical curve for typical collector panel efficiency versus ($T_i - T_a$)/I:

$$\frac{\text{mean collector temp.} - \text{ambient temp.}}{\text{solar flux}}$$

curve is represented as a straight line with an intercept of 0.78 and a slope of -0.65.

$$\eta = 0.78 - 0.65\nu \qquad \text{(7-5)}$$

where: η = efficiency
ν = operating parameter.

In a forced circulation system, the average fluid temperature may be selected by subtracting 5 or 10 degrees from the required service temperature. In this example, required service temperature is 145°F so that average fluid temperature circulating through the collector would be $\overline{T}_F = 140°F$. A summary of all fluid temperatures in a system operating on a clear, cloudless day would be as follows: fluid entering the collector: $T_{in} = 135°F$, water exiting the collector after solar heating: $T_{out} = 145°F$, average fluid temperature in the collector $\overline{T}_F = 140°F$. Note that the use of a heat exchanger between collector and storage, a common system component because of the advantages it yields, also can degrade system performance slightly because of thermal losses in the heat exchanger. Thus, in this example, T_{out} of 145°F exiting the collector and entering one loop of the heat exchanger will yield fluid temperatures of 135°–140°F in

TABLE 7-10 BTUs Delivered Worksheet

City: Philadelphia
Latitude: 40°
Longitude: 75°
Elevation: 200 ft
Service temperature: 145°F

Collector tilt angles:
From horizontal: $\Sigma = 45°$
From due south: $\phi = +15°$
Reflection coefficient: $C_r = 0.25$
Average plate temperature: $\overline{T}_F = 140°F$

Solar Time	Jan. ν	Jan. η	Jan. BTU DEL.	Feb. ν	Feb. η	Feb. BTU DEL.	Mar. ν	Mar. η	Mar. BTU DEL.	Apr. ν	Apr. η	Apr. BTU DEL.	May ν	May η	May BTU DEL.	June ν	June η	June BTU DEL.
6:00										4.14	—	0	2.66	—	0	2.34	—	0
7:00				4.61	—	0	1.26	—	0	0.86	.22	22	0.77	.28	28	0.70	.32	31
8:00	1.11	.06	5	0.71	.32	47	0.54	.43	78	0.48	.47	85	0.44	.50	87	0.41	.51	85
9:00	0.51	.45	94	0.44	.50	120	0.38	.53	138	0.35	.55	139	0.33	.57	134	0.30	.58	131
10:00	0.39	.52	144	0.36	.55	163	0.31	.58	180	0.29	.59	178	0.28	.60	168	0.26	.61	163
11:00	0.36	.55	166	0.32	.57	187	0.29	.59	199	0.27	.61	196	0.26	.61	184	0.24	.63	179
12:00	0.36	.55	165	0.33	.57	185	0.29	.59	198	0.27	.60	195	0.26	.61	183	0.24	.62	176
1:00	0.39	.52	144	0.36	.55	160	0.32	.57	175	0.30	.59	173	0.28	.60	161	0.26	.61	157
2:00	0.50	.46	100	0.44	.50	120	0.39	.53	133	0.36	.55	133	0.34	.56	126	0.32	.57	123
3:00	0.77	.28	40	0.63	.37	61	0.56	.42	74	0.50	.45	78	0.47	.47	77	0.44	.49	75
4:00	2.16	—	0	1.39	—	0	1.07	.09	8	0.94	.17	16	0.86	.22	20	0.81	.25	21
5:00							5.16	—	0	3.00	—	0	2.41	—	0	2.13	—	0
6:00																		
T_A		32			34			42			53			63			72	
$\overline{T}_F - T_A$		108			106			98			87			77			68	

BTUs delivered per day per square foot

	Jan.	Feb.	Mar.	Apr.	May	June
Clear day	858	1043	1183	1215	1168	1141
Per month	26,600	29,200	36,700	36,500	36,200	34,200
× % sun	13,300	15,500	20,900	20,400	21,000	21,600

the service loop. These temperature figures are only approximate, for purposes of example, and a well-designed system operating under good weather conditions would achieve higher temperatures. The ambient temperature for the design purposes of the example is assumed to be the average ambient temperature from the Site Climate Worksheet with total solar insolation obtained from the Total Insolation Worksheet. The BTUs delivered is found by simply multiplying the total insolation (from the worksheet of Step 3) by the efficiency of the collector; these data are entered in the BTUs Delivered Worksheet (Table 7–10) which summarizes this information, operating parameter, and collector efficiency.

STEP 6

Compare the BTUs delivered per square foot with the service load requirement. The service load requirements and BTU requirements are approximately calculated on a monthly basis as follows:

Hot Water. The number of baths and bedrooms used to determine the BTU requirements in the example are taken from Table 7–1. Given a 2½ bathroom, 4

TABLE 7–10 BTUs Delivered Worksheet (continued)

City: Philadelphia
Latitude: 40°
Longitude: 75°
Elevation: 200 ft
Service temperature: 145°F

Collector tilt angles:
  From horizontal: $\Sigma = 45°$
  From due south: $\phi = +15°$
  Reflection coefficient: $C_r = 0.25$
  Average plate temperature: $\bar{T}_F = 140°F$

Solar Time	July ν	July η	July BTU DEL.	Aug. ν	Aug. η	Aug. BTU DEL.	Sept. ν	Sept. η	Sept. BTU DEL.	Oct. ν	Oct. η	Oct. BTU DEL.	Nov. ν	Nov. η	Nov. BTU DEL.	Dec. ν	Dec. η	Dec. BTU DEL.
6:00	2.42	—	0	3.82	—	0												
7:00	0.68	.34	32	0.73	.31	27	1.09	.07	5	4.88	—	0						
8:00	0.38	.53	89	0.39	.53	87	0.44	.49	81	0.61	.38	51	1.04	.10	9	1.98	—	0
9:00	0.28	.60	135	0.28	.60	139	0.31	.58	137	0.37	.54	122	0.46	.48	97	0.63	.37	61
10:00	0.24	.63	167	0.23	.63	176	0.25	.62	179	0.30	.59	164	0.35	.55	147	0.41	.52	134
11:00	0.22	.64	184	0.22	.64	193	0.23	.63	197	0.27	.61	188	0.32	.57	170	0.36	.55	159
12:00	0.22	.64	182	0.22	.64	193	0.23	.63	197	0.27	.60	186	0.32	.57	170	0.36	.55	162
1:00	0.24	.62	162	0.24	.63	171	0.25	.62	175	0.30	.59	163	0.35	.55	148	0.40	.52	137
2:00	0.29	.59	127	0.29	.59	133	0.31	.58	135	0.37	.54	123	0.44	.49	105	0.50	.45	95
3:00	0.41	.51	79	0.41	.51	82	0.45	.49	79	0.53	.43	68	0.68	.34	47	0.81	.25	32
4:00	0.75	.29	25	0.78	.27	22	0.88	.21	17	1.20	0	0	1.96	—	0	3.75	—	0
5:00	2.17	—	0	2.71	—	0	4.50	—	0									
6:00																		
T_A		77			75			68			57			46			35	
$\bar{T}_F - T_A$		63			65			72			83			94			105	

BTUs delivered per day per square foot						
Clear day	1182	1223	1202	1065	893	780
Per month	36,600	37,900	36,100	33,000	26,800	24,200
× % sun	22,700	23,900	21,600	19,500	13,900	12,100

TABLE 7–11 Hot Water Worksheet

City: Philadelphia

Collector angles: $\phi = 15°$
$\Psi = 45°$
$\Sigma = 45°$

Reflection coefficient: $C_r = 0.25$

House: 2000 ft², insulated,
2½ baths, 4 bedrooms

Collector: AMETEK, double-glazed

	Jan.	Feb.	Mar.	Apr.	May	June	July	Aug.	Sept.	Oct.	Nov.	Dec.	Average All Year
Hot water BTUs received (10^6 BTU/mo)	1.2	1.1	1.2	1.1	1.2	1.1	1.2	1.2	1.1	1.2	1.1	1.2	1.1
BTUs delivered per ft² (1000 BTU/ft²)	6	11	18	20	24	26	27	27	24	20	12	7	18.5
Ratio (ft²)	200	100	67	55	50	42	44	44	46	60	92	171	—
% of needs met by 3 AMETEK collectors (48 ft²)	24	48	72	87	96	100	100	100	100	80	52	28	74
% of needs met by 6 AMETEK collectors (96 ft²)	48	96	100	100	100	100	100	100	100	100	100	56	92

TABLE 7–12 Space Heating Worksheet

City: Philadelphia

Collector angles: $\phi = 15°$
$\Psi = 45°$
$\Sigma = 45°$

Reflection coefficient: $C_r = 0.25$

House: 2000 ft², insulated,
2½ baths, 4 bedrooms

Collector: AMETEK, double-glazed

	Jan.	Feb.	Mar.	Apr.	May	June	July	Aug.	Sept.	Oct.	Nov.	Dec.	Average All Year
Degree days	1000	870	770	370	120	0	0	0	40	250	560	920	408
BTU heating requirement	20	17	14	7.4	2.4	0	0	0	0.8	5.0	11	18	8.0
BTUs delivered per ft² of collector	6	11	18	20	24	26	27	27	24	20	12	7	18.5
Ratio (ft²)	3300	1500	780	370	100	0	0	0	33	250	920	2600	—
% of needs met by 30 AMETEK collectors (480 ft²)	15	32	62	100	100	100	100	100	100	100	52	18	73
% of needs met by 40 AMETEK collectors (640 ft²)	19	43	82	100	100	100	100	100	100	100	70	25	70
% of needs met by 60 AMETEK collectors (960 ft²)	29	64	100	100	100	100	100	100	100	100	100	37	86

bedroom house, the minimum requirement would be 38,000 BTU per day. For hot water service of 145°F, the BTUs Delivered Worksheet provides the monthly BTUs per square foot of collector. The collector area may be selected to supply 100% of the BTU requirements in the coldest average month but this will represent the largest area to be considered. It will be more economical to use a smaller collector area and expect to use back-up heating during the coldest months. The ratio of BTUs required per month to BTUs delivered per square foot per month gives the area necessary to meet 100% of the energy requirements in that month.

Space Heating. The monthly space heating BTU requirement should be calculated or measured according to the ASHRAE methods outlined earlier,

TABLE 7–13 Air Conditioning Worksheet

City: Philadelphia

Collector Angles: $\phi = 15°$
$\Psi = 45°$
$\Sigma = 45°$

Reflection coefficient: $C_r = 0.25$

House: 2000 ft², insulated, 2½ baths, 4 bedrooms

Collector: AMETEK, double-glazed

	Jan.	Feb.	Mar.	Apr.	May	June	July	Aug.	Sept.	Oct.	Nov.	Dec.	Average All Year
Normal maximum temperature	40	42	51	64	74	83	87	85	78	68	56	43	64
Degrees exceeding 65°F	—	—	—	—	9	18	22	20	13	3	—	—	—
Days per month	31	28	31	30	31	30	31	31	30	31	30	31	—
Cooling degree days	0	0	0	0	279	540	682	620	390	93	0	0	—
Relative humidity	.59	.57	.53	.49	.52	.54	.55	.54	.56	.53	.56	.61	.55
A/C load (10⁶ BTU/mo)	0	0	0	0	8.3	16.7	21.4	19.1	12.5	2.8	0	0	—
A/C BTUs delivered (1000 BTU/ mo ft²)	—	—	—	—	17.4	19.2	20.0	20.3	17.6	14.7	—	—	—
Ratio (ft²)	—	—	—	—	480	870	1070	940	710	190	—	—	—
% of needs met by 30 AMETEK collectors (480 ft²)	—	—	—	—	100	55	45	51	68	100	—	—	—
% of needs met by 40 AMETEK collectors (640 ft²)	—	—	—	—	100	74	60	68	90	100	—	—	—
% of needs met by 60 AMETEK collectors (960 ft²)	—	—	—	—	100	100	90	100	100	100	—	—	—

but a quick computation may be gained by using the following approximate technique:

BTU heating requirement =

$F \times$ (floor area in square feet) $\times$ (monthly degree days)

where: $F = 12$ for a poorly insulated house

$F = 10$ for an insulated house

$F = \ \ 8$ for a well-insulated house.

In the example, $F = 10$ and a 2000-square-foot floor area is given in the determination of the monthly BTU requirements for space heating.

Air Conditioning. At the present time, air conditioning requires a service temperature of approximately 190°F. The lithium bromide absorption system is the most efficient air-conditioning technique with input energy available from thermal sources now known. Alternatively, heat pumps may use mechanical energy to drive a compressor, but the efficiency of conversion of solar heated fluids to mechanical energy is poor without the use of concentrating collectors.

For those air-conditioning applications with a 200°F temperature requirement, BTUs delivered must be recomputed for a 200 degree service during the summer months, as has been done on the Air Conditioning BTUs Delivered Worksheet. The value of T_F used in the calculation is $200° - 5° = 195°F$.

In practice, air conditioning loads should be determined using the AS-HRAE procedure, since the load depends on humidity as well as outside temperature. An approximate method, useful for estimating air conditioning load, is:

$$AC \ load = F \ \frac{2 \times (\text{relative humidity})}{COP} \ (\text{floor area, ft}^2) \ (\text{monthly cooling degree days})$$

TABLE 7–14   Collector Area Worksheet

City: Philadelphia

Collector angles: $\phi = 15°$

$\Psi = 45°$

$\Sigma = 45°$

Reflection coefficient: $C_r = 0.25$

House: 2000 ft², insulated

2½ baths, 4 bedrooms

Collector: AMETEK, double-glazed

Number of Collectors	Area (ft²)		Percent of Needs Met During the Year		
	Effective Solar Collection	Physical*	Hot Water	Space Heating	Air Conditioning
3	48	54	74	—	—
6	96	108	92	—	—
30	480	540	100	53	60
40	640	720	100	65	75
60	960	1080	100	82	97

* The physical area is the area of the collector, including its box or frame. Additional area is usually required for fluid connection headers and access paths for maintenance.

where: *AC* load, is the air conditioning BTU hot water requirement

$$F = \begin{cases} 12 \text{ for poorly insulated house} \\ 10 \text{ for insulated house} \\ 8 \text{ for well-insulated house} \end{cases}$$

COP is the coefficient of performance for the air-conditioning system used.

The monthly cooling degree days are obtained by multiplying the number of days in a month by the amount by which the normal monthly average temperature (Table 3–5) exceeds 65°F. The average percent humidity data necessary for the calculation is contained in Table 3–4. In the example, based on a site near Philadelphia with a 2000-square-foot home with $F = 10$ and an absorption unit with $COP = 0.7$, the relevant figures are contained in the Air Conditioning Worksheet (Table 7–13).

TABLE 7–15 Solar Storage Capacity Factors (Gallons Per Square Foot of Collector Area)

Sunshine	Average Annual Percent Sun (from Table 3–1)	Storage Factor (Gal/ft²)		
Cloudy	0–55 ⟶	1.9	1.7	1.5
Normal	56–69 ⟶	1.7	1.5	1.3
Sunny	70–100 ⟶	1.5	1.3	1.1

Temperature Extremes	Difference Between Annual Extremes (from Tables 3–6, 3–7)
High	120–150
Normal	90–120
Low	0–90

1.9	1.7	1.5
Cleveland Detroit	Seattle Portland, OR Wilmington, DE	

1.7	1.5	1.3
New York Chicago Washington, DC Omaha Salt Lake City Portland, ME Minneapolis Nashville	Boston Philadelphia Providence Baltimore Atlanta Houston Richmond Charlotte	Miami San Francisco Honolulu

1.5	1.3	1.1
Denver Albuquerque	Phoenix Reno	Los Angeles

The number of solar collectors needed is summarized in the Collector Area Worksheet (Table 7–14), which consolidates the information obtained on previous worksheets.

STEP 7

Determination of storage capacity. The storage capacity for a solar heating system should range from one to two gallons for each square foot of collection area. If it is less than one gallon, it will be in danger of overheating; storage of more than two gallons per square foot of collector may result in lower storage temperature levels than are desirable during any one day.

Storage capacity for hot water heating may be calculated at 1.5 gallons per square foot. In the example, using three AMETEK collectors, this amounts to 72 gallons (1.5 gal/ft^2 × 48 ft^2); the use of six AMETEK collectors would require 144 gallons of storage capacity.

The three-collector example supplies 74% of the total hot water heating BTU requirement with fulfillment of the remaining 26% to be achieved with a back-up water heater. In this example, two 40-gallon insulated hot water tanks with heat exchange loops would provide adequate service. The use of two exchange loops is necessary not only because of building code demands, but to prevent contamination of drinking water by the collector antifreeze should leaks develop inside either tank. The use of two storage tanks has other advantages as well. It allows for safe temperature control through control of the intertank circulation pump without disturbing the collector circulation pump. It also allows for large separation distances between storage tanks to take advantage of bulk transfer between the tanks which reduces the thermal loss in the connecting pipes.

In the six-collector example, the solar storage tank will require a capacity of about 100 gallons, but it will supply 92% of hot water heating for the location and tilt angles used.

In space heating, the storage requirements are also calculated at 1.5 gallons per square foot of collector area, while for air conditioning it may be more desirable to use 1.2 gallons per square foot for more rapid achievement of the high storage water temperatures necessary for the supply of the absorption apparatus. The factors that would be used to determine the storage capacity requirements are listed in Table 7–15.

CHAPTER 8
ECONOMIC ANALYSIS

Economic analysis can play a vital role in solar energy system planning and implementation procedures. In the planning of a new system it provides a means of determining the cost effectiveness of the various utilization techniques under consideration. With an existing system it can determine whether the performance advantages of higher priced equipment, if added to a facility, are worth the added cost.

The process is a familiar one to every consumer who determines the economic justification of a major expenditure by comparing the continuing benefits derived from it to the monthly or annual payments required by it. This comparison is the foundation of the approach employed by Richard M. Winegarner as published in the *Optical Industry and Systems Directory*, 1977, in which the justifiable costs of solar energy equipment are based upon the savings to be derived. The direct economic benefits produced by a solar energy system can be determined from the product of the insolation rate, efficiency, and cost of conventional energy. The cost per period or payments required to operate a solar energy system are simply a product of the amortization rate and the cost per unit area of the installed system. There are other factors, such as the beneficial effects on the environment and improvement in the nation's balance of payments which could stimulate even earlier use through governmental incentives, but this analysis considers only direct benefits. In sum, any solar energy system becomes viable whenever the value of the energy produced exceeds the payments required.

A SIMPLIFIED ANALYSIS

Since the cost level for flat plate collectors is directly related to the product of the yearly insolation rate and the conventional utility rate, economic analysis can be performed by equating the economic benefits and the payments required:

$$I\eta_c\eta_d E = RC \qquad \text{(8–1)}$$

where: I = insolation rate in MBTU/yr/ft²

η_c = percentage collector efficiency

η_d = distribution efficiency

E = conventional energy cost \$/MBTU

R = percentage amortization rate

C = installed solar energy system cost in \$/ft².

Representative values of I and E are shown in Table 8–1.

TABLE 8–1 Insolation Rate: Electric Rate Product at Several U.S. Locations

Location	I MBTU/Yr/Ft^2	E $/MTBU	$I \times E$ $/Yr/Ft^2
Northeast	.47	12	5.6
Southern California	.65	8	5.2
Tennessee	.51	7	3.6
Texas	.58	6	3.5
Northwest	.36	3	1.1

Solving the equation for cost per square foot:

$$C = \frac{\eta_c \eta_d IE}{R} \tag{8–2}$$

Differentiating this equation with respect to collector efficiency yields:

$$\Delta C = \frac{\eta_d IE \, \Delta \eta_c}{R} \tag{8–3}$$

With this equation we can perform marginal analysis on any system to determine whether the addition or deletion of certain components makes a system more or less economically viable.

As an example, consider the case of a residential electrical heating and cooling system augmented by a flat plate solar energy collection system. Assume that the addition of a heat mirror coating placed on the third cover surface of a standard flat plate collector is being considered. Will the benefits derived from the addition of the mirror coating more than offset the additional costs?

The average residential customer price for electricity in the United States in 1974 was $8.78/MBTU; suppose that addition of the heat mirror coating will produce an increase of 10% in collector efficiency. With a 20-year system lifetime and a money rate of 8.5%, the amortization rate would be approximately 10%. Using these numbers with an average insolation rate of 0.5 MBTU per year per square foot and a distribution efficiency of 0.85, the marginal costs can be determined as follows:

$$\Delta C = \frac{0.85 \times 0.5 \text{MBTU/yr/ft}^2 \times \$8.78/\text{MBTU}}{10\%/\text{yr}} \; 9.9\% = \$3.69/\text{ft}^2$$

If the addition of the coating allows the solar system to collect additional energy valued at $3.69 per square foot, the system will be more economically viable. Similar analyses can be performed to determine the economic feasibility of many controversial items, such as the number of covers, cover materials (glass or plastics), honeycombs, selective absorber surfaces, and so on. Care must be taken when analyzing a concentrator system, especially when the modification is made only to a reduced area within the system, to include the ratio of the modified or coated area to the total area in the equation as follows:

$$\Delta C = \frac{\eta_e IE}{R} \frac{A_t}{A_m} \Delta \eta_c \tag{8–4}$$

where: A_t = total area of collector

A_m = reduced area modified

η_e = electrical conversion efficiency.

The addition of a selective absorber to a trough solar collector with a concentration ratio of 12 to 1 under the same economic conditions would produce the following results, assuming an electrical conversion efficiency of 0.36:

$$\Delta C = \frac{0.36 \times 0.5\text{MBTU/yr/ft}^2 \times \$8.78\text{/MBTU}}{10\%\text{/yr}} \frac{12}{\pi} \, 22.5\%$$

$$= \$13.58 \text{ per square foot}$$

Thus, if a selective absorber coating can be applied to the receiver tube of a concentrating trough collector at a cost less than $13.58 per square foot (or, stated in another way, will enable the collector to capture additional energy of equivalent value per square foot), the equipment alteration becomes economically worthwhile.

COMPOUND INTEREST AND GROWTH OF INCOME

Since the income produced by an asset consisting of solar energy equipment is equated to the cost of energy saved from conventional sources, it is useful to consider the present value of an asset which produces an income that is increasing at a steady rate of growth.
Consider an asset which produces an income given by:

$$M_k = Me^{(k-1)r_g} \qquad (8\text{--}5)$$

Where M_k is the income produced in the kth year (or period), k is the year (or period) number and r_g is the growth rate (% ÷ 100) of the income stream. This is representative of an asset which replaces some cost which is growing at a rate, r_g, due to inflation, shortages, or increasing production costs. The present value of the asset, assuming that it has a life of n years (or periods) is:

$$PV = Me^{-r_g} \left[\frac{e^{n(r_g - r)} - 1}{1 - e^{-(r_g - r)}} \right] - e^{-nr}(S_r - S_s) \qquad (8\text{--}6)$$

Where r is the nominal interest rate per year (or period), M is the cost saving in the first year (or period), $e = 2.71828$, r_g is the rate of growth of cost saving per year (or period), $r_g \neq r$, n is the lifetime of the asset in years (or periods) and S_r and S_s are the replacement cost and salvage value of the asset, respectively, after n years (or periods). Equation 8–6 is based on continuous compounding of interest.

If $r_g = r$, a limiting form for Equation 8–6 becomes:

$$PV = nMe^{-r} - e^{-rn}(S_r - S_s) \qquad (8\text{--}7)$$

The value of a solar energy system can be analyzed in terms of its financial value, using Equation 8–6. It is instructive to compare the present value of a solar system as determined by Equation 8–6 to its cost. When and if the cost should fall below the present value, then it makes economic sense to purchase a solar energy system.

The examples that follow are used to illustrate certain characteristics of solar energy systems as an investment.

Example 1: Consider a solar energy system that will initially save $500 per year in conventional fuel cost. Assume that fuel costs are expected to increase at a uniform rate of 12% per year and that the solar system can be financed through an extension of a home mortgage loan with a nominal interest rate of 8%. Assume, also, that the solar system has a design life of 20 years and a salvage value equal to its replacement cost (it is possible that the salvage value could exceed the initial cost, depending on the system condition, inflation in raw materials, new technologies available, and numerous other factors). Using Equation 8–6:

$$M = 500$$
$$r_g = 0.12$$
$$r = 0.08$$
$$n = 20$$
$$S_r - S_s = 0$$

and calculate $PV = \$13,860.25$ from Equation 8–6.

Example 2: Consider the same conditions as in Example 1, except that the salvage value is zero and the replacement cost is $10,000. Then:

$$S_r - S_s = \$10,000$$

and from Equation 8–6:

$$PV = \$11,841.29$$

Example 3: Consider the same conditions as in Example 1, except that the design life is 30 years instead of 20. Then $n = 30$ and from Equation 8–6:

$$PV = \$26,240.03$$

Example 4: Consider the same conditions as in Example 3, except that the difference between replacement cost and salvage value is $10,000 so that $S_r - S_s = -10,000$ and

$$PV = \$25,332.85$$

Note that the present value rises significantly as the design life of the solar energy system is extended, giving greater value per unit of service time as the design life is increased.

Example 5: Consider a severe energy crisis scenario where the growth rate of conventional fuel cost is 20% ($r_g = 0.20$) but where government backed incentives and subsidies provide low-interest 5% loans to solar energy system purchases so that $r = 0.05$. Let the initial savings be $500 per year and consider both 20 and 30-year design life, with replacement cost exceeding salvage value by $10,000:

20-year design life, $PV = \$52,410.48$
30-year design life, $PV = \$259,368.77$

Clearly, in this scenario, a solar system could be considered highly economical, especially if it has long design life.

Example 6: Consider an economic depression scenario where reduced use of conventional fuels and new oil and gas discoveries result in a modest 5% rate of growth for fuel costs ($r_g = 0.05$) and assume interest rates are 6%, or $r = 0.06$. Otherwise, the same conditions as in Example 5 would prevail:

20-year design life, PV = $5,611.05
30-year design life, PV = $10,675.08

Examples 5 and 6 are both extreme, perhaps improbable, but they serve to illustrate the effects of compounding on interest costs and on the growth of alternative energy costs.

The decision to invest in a solar energy system on a financial basis alone rests on assumptions that extend far into the future. Provided that the design life of the equipment is adequately long and using the initial cost savings of Table 3–14, with reasonable assumptions concerning interest rates and fuel cost growth rates, it becomes readily apparent that the costs of a solar energy system shown in Table 3–16 compare favorably with the present value of such a system. It is important to note that economy in costs, both initial and operating, is a function of equipment quality. Cheaper system components, particularly solar collection panels, may seem more attractive because of initial price. But they are low in price for a reason which, too often, translates into construction shortcuts that result in lower efficiencies and shortened equipment life. It is for this reason that the "disposable quality" or "planned obsolescence" characteristics of many products cannot be justified in solar energy equipment. It will not be economical unless it is built to last a long time.

Another way to understand the importance of quality in solar energy equipment is to consider that the energy expended in the production of a solar energy system (including all steps, from the mining and processing of the natural resources used to make it, through its final assembly and installation) must be more than replenished by the energy it collects during its lifetime. Otherwise, the equipment is only contributing to a net drain on total energy resources. Thus, quality and endurance are not only most important features of solar energy equipment, but also the twin keys to achieving the potential of superior long-term economy and contributions to improvement of the nation's energy shortage.

MINIMUM COST ANALYSIS

The concept of minimizing the present value equivalent of future costs is used in determining the optimum collector area for a solar energy system. An example of this analysis is shown in Figure 8–1, where it is assumed that the present value of solar system cost is $33 per square foot of collection area. This figure portrays the sum of the initial solar energy cost and the present value of its continuing maintenance costs as a linear function of collector area. For a 1500-square-foot home in the Washington, D.C., area, the present value of the future costs of conventional fuel necessary to supplement energy requirements is shown by the declining curve on the figure. Note that the shape of this curve is sensitive to assumptions about the future cost of fuels and other alternatives for home heating. The uppermost curve represents the sum of equipment cost

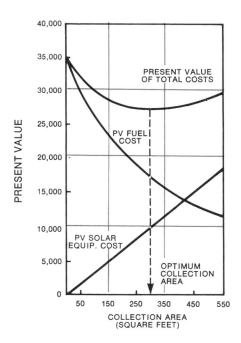

FIGURE 8–1 Optimization of collection area: Washington, D.C., 1500-square-foot home.

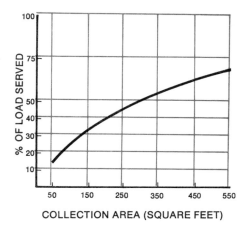

FIGURE 8-2 Percent of annual hot water and heating load as a function of collection area: Washington, D.C., 1500-square-foot home.

and fuel costs—in a phrase, the present value of total system costs. The minimum point of the total costs curve, marked by the dotted line, is used to determine the optimum collection area—optimum in the sense that the present value of total costs is minimum.

Although the example represented in Figure 8–1 applies to a specific situation with many assumptions, the results are qualitatively similar in most solar energy applications. A similar plotting for any application will show that it is most economical to design a solar energy system to provide approximately 50%–80% of total energy demand.

Figure 8–2 illustrates the relationship between the percent of load served and the collection area for the Washington, D.C., home of 1500 square feet. It can be seen from the curve, called an f-chart and applicable only to the example under discussion, that the acquisition of additional collection area results in a diminishing increment of improvement in the percent of load served. An iterative computer program was developed at the University of Wisconsin to produce an f-chart for engineering specific solar energy systems; the worksheets in Chapter 7 of this handbook can be used to yield a similar evaluation by determining the relationships between percent of load served and collection area on a point-by-point basis for a wide variety of collector orientations and climatic conditions. The curve shown in Figure 8–2, however, is qualitatively similar to that which could be expected anywhere.

APPENDIX A
BLACKBODY RADIATION
LAWS

A blackbody, so called because of the resulting color, is defined as an object that absorbs all and reflects none of the radiation incident upon it. No perfect blackbody exists but the concept is important because the temperature of a blackbody can be mathematically related to the proportion between its radiant absorption and emission and, hence, a relation to the properties of real objects in actual radiative environments.

A blackbody emits radiation in accordance with Planck's Law:

$$S = \frac{C_1}{\lambda^5(e^{C_2/\lambda T} - 1)}$$
(A–1)

where: S is the spectral radiation intensity in watts per square meter of area per meter of wavelength, emitted from the surface of the blackbody

T is the absolute temperature of the blackbody in °K

λ is the wavelength of the radiation emitted, in meters

C_1 is the first radiation constant
$\quad C_1 = 2\pi hc^2 = 3.7418 \times 10^{-16}$ Joule-meter/second

C_2 is the second radiation constant
$\quad C_2 = \frac{hc}{k} = 0.014388$ meter – °K

and where: $h = 6.6262 \times 10^{-34}$ Joule-seconds is Planck's constant

$c = 2.998 \times 10^8$ meters/second is the speed of light

$k = 1.3806 \times 10^{-23}$ Joules/°K is Boltzmann's constant.

Interestingly, as shown in Figure 2–1, the application of Planck's Law to a blackbody with an absolute temperature of 5750 K, near that of the solar surface, results in a theoretical curve that closely approximates the curve for the sun. Figure 2–1 also compares the theoretical characteristics of both the sun and the earth to their estimated actual emissions.

The total power emitted by a blackbody is found by integrating the spectral radiation intensity, S, given by Equation A–1, over all values of wavelength, λ. The result provides a relationship between the radiation intensity in units of power emitted per unit area, and the absolute temperature of a blackbody. This relationship is the Stefan-Boltzmann equation, more commonly known as the "fourth power law":

$$I = \sigma T^4$$
(A–2)

where: I is the radiation intensity, or power emitted per unit area from a blackbody at temperature T

T is the absolute temperature in °K

$\sigma = \dfrac{2\pi^5 k^4}{15h^3 c^2} = 5.6996 \times 10^{-6}$ watts/(meter^2 · °K^4) is the Stefan-Boltzmann constant.

One convenient expression for the radiation intensity of a blackbody is obtained when T is in °K and I is in watts/meter^2:

$$I = \left(\frac{T}{64.8}\right)^4 \qquad \text{watts/meter}^2 \tag{A-3}$$

The wavelength at which the maximum spectral radiation intensity occurs is given by Wien's Displacement Law:

$$\lambda_{max} = \frac{2.898 \times 10^{-3}}{T} \qquad \text{meters} \tag{A-4}$$

Wien's Displacement Law is a quantitative expression to describe the displacement of peak radiation intensities toward shorter wavelengths as an object is heated. This displacement is readily seen in Figure 2–1 by comparing blackbody radiation at 5750°K and 300°K. Wien's Law also provides a basis for real objects to radiate maximum energy at different wavelengths than those at which energy is best absorbed. There are several useful effects which result from Wien's Law in the application of materials with selective absorption and reflectance of energy in different wavelength regions. One of the best known is the "greenhouse effect," which is based upon the property of glass in transmitting light but not radiant heat. Consequently, the interior of a greenhouse can be kept warm on sunny winter days as light energy is allowed to enter, but the radiant heat wavelengths are displaced by Wien's Law so that they cannot escape by direct reradiation. Another useful effect is based on selective absorption and emission by certain surfaces. These selective surfaces, discussed in detail in Chapter 4, have the property of absorbing light efficiently and converting it to heat for which emissivity is poor at the displaced wavelengths described by Wien's Law.

Absorptivity, emissivity, transmission, and reflection coefficients may be used directly in conjunction with the blackbody radiation laws to derive practical formulas for the radiation energy gained or lost by a real object in terms of its temperature. Conversely, by using the principle of equilibrium (equating energy gained with energy lost), one can determine the temperature of a real object in a radiative environment.

A real object emits radiation at a rate given by:

$$I_{real} = \epsilon \left(\frac{T}{64.8}\right)^4 \tag{A-5}$$

where ϵ is the emissivity of the object and T is the absolute temperature of the object in °K. The spectral composition of the emitted radiation is given by:

$$S_{real} = \epsilon S \tag{A-6}$$

where S is the blackbody spectral radiation intensity in Equation A–1 and the magnitude of maximum wavelength is given by Wien's Law if the emissivity,

ϵ, is constant in the vicinity of that maximum wavelength. (Remember that ϵ may itself be a function of λ, T, and the incident angle.)

An example of the use of these radiative laws is to determine the temperature of an object in a vacuum that is exposed to the sun. It will gain energy at the rate of αI where α is the absorptivity of the object and I is the intensity of solar energy incident upon it. It will lose energy only by reradiating its gained energy in the form of heat to its surroundings at the same rate once its temperature has stabilized.

Therefore:

$$\alpha I = \epsilon \left(\frac{T_{object}}{64.8} \right)^4 \tag{A-7}$$

or

$$T_{object} = 64.8 \; \sqrt[4]{\frac{\alpha}{\epsilon} I} \tag{A-8}$$

It must be assumed that the object is not absorbing radiant heat energy from its surroundings.

As noted in Chapter 4, covers should be constructed of materials which possess a high transmittance, τ, for solar radiation, low reflectance, ρ, low absorptance, α, and, of course, should be able to withstand severe environmental hazards, such as the impact of hailstones and high winds. The reflectance, ρ, and transmittance, τ, of a cover plate can be determined from the surface reflectance, ρ_s, absorptance, α, and thickness, x, of the cover plates. The surface reflectance, ρ_s, is determined from the index of refraction, n, the angle of incidence, θ, and the polarization of the incident radiation. For unpolarized light:

$$\rho_s = \frac{1}{2} \left(\frac{\sin^2 (\theta - \psi)}{\sin^2 (\theta + \psi)} + \frac{\tan^2 (\theta - \psi)}{\tan^2 (\theta + \psi)} \right) \tag{B-1}$$
$$\text{(Fresnel's formula)}$$

where: $\sin \psi = (\sin \theta)/n$ (Snell's Law)
determines the angle of refraction, ψ, where for air, $n = 1$.

The transmittance, reflectance, and absorptance are affected by internal reflections which are infinitely summed in the following closed expressions:

$$\alpha = \frac{\alpha_0 \, x}{\cos \psi} \, * \tag{B-2}$$

$$\tau = \frac{(1 - \rho_s)^2 \, (1 - \alpha)}{1 - (1 - \alpha)^2 \, \rho_s^2} \tag{B-3}$$

$$\rho = \rho_s \left[1 + \frac{(1 - \rho_s)^2 \, (1 - \alpha)^2}{1 - \rho_s^2 \, (1 - \alpha)^2} \right] \tag{B-4}$$

Both the transmitted and reflected light will be slightly polarized for $\theta \neq 0$.

By ignoring the effects of polarization, the compound transmittance, $\tau_{1,2}$, reflectance, $\rho_{1,2}$, and absorptance, $\alpha_{1,2}$, of two covers with generally different

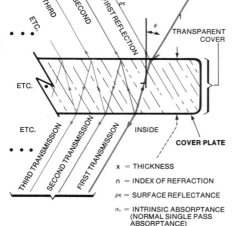

x = THICKNESS

n = INDEX OF REFRACTION

ρ_s = SURFACE REFLECTANCE

α_o = INTRINSIC ABSORPTANCE (NORMAL SINGLE PASS ABSORPTANCE)

FIGURE B-1 Internal reflections in transparent cover.

---

* α may be expressed more precisely by Bouger's Law as follows:

$$\alpha = 1 - e^{\frac{\alpha_0 x}{\cos \psi}}$$

properties are determined as:

$$\tau_{1,2} = \frac{\tau_1 \tau_2}{1 - \rho_1 \rho_2} \qquad \textbf{(B–5)}$$

$$\rho_{1,2} = \rho_1 + \frac{\tau_1^2 \rho_2}{1 - \rho_1 \rho_2} \qquad \textbf{(B–6)}$$

$$\alpha_{1,2} = 1 - \tau_{1,2} - \rho_{1,2} \qquad \textbf{(B–7)}$$

For three dissimilar covers, the transmittance and reflectance are given by:

$$\tau_{1,2,3} = \frac{\tau_{1,2} \tau_3}{1 - \rho_{1,2} \rho_3} \qquad \textbf{(B–8)}$$

$$\rho_{1,2,3} = \rho_{1,2} + \frac{\tau_{1,2}^2 \rho_3}{\rho_{2,1} \rho_3} \qquad \textbf{(B–9)}$$

where: $\tau_{1,2,3}$ is the transmittance of three covers

$\rho_{1,2,3}$ is the reflectance of three covers

$\tau_{1,2}$ is the transmittance of two covers given by Equation B–5

$\rho_{1,2}$ is the reflectance of two covers given by Equation B–6

$\rho_{2,1}$ is the reflectance of two covers from the opposite side given by interchanging subscripts in Equation B–6 ($\rho_{1,2} \neq \rho_{2,1}$ for dissimilar covers)

τ_3 is the transmittance of the third cover given by Equation B–3

ρ_3 is the reflectance of the third cover given by Equation B–4.

Equations of the form shown in B–8 and B–9 may be extended to any number of covers by adding one cover at a time and by giving proper attention to subscript order.

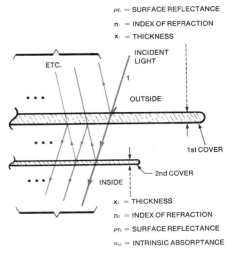

α_{o_i} = INTRINSIC ABSORPTANCE
ρs_i = SURFACE REFLECTANCE
n_i = INDEX OF REFRACTION
x_i = THICKNESS

INCIDENT LIGHT

ETC.

1

OUTSIDE

1st COVER

2nd COVER

INSIDE

x_2 = THICKNESS
n_2 = INDEX OF REFRACTION
ρs_2 = SURFACE REFLECTANCE
α_{o_2} = INTRINSIC ABSORPTANCE

FIGURE B-2 Internal reflections in multiple covers.

LIMITING ACCURACIES

The limiting value of measurement uncertainty attainable under laboratory conditions for basic physical measurements has been reported by the United States Bureau of Standards. A summary of these limiting accuracies is provided in Table C–1 in those ranges of interest for flat plate solar collectors.

TABLE C–1   Limiting Accuracies of Measurement

Parameter	Range of Interest	Technique	Measurement Uncertainty One Part I_n (%)
Temperature	200°–500°K	Platinum resistance thermometry	$10^5 = 0.001$
Liquid mass flow	10–1000 lb/hr	Density corrected volume flow	$10^3 = 0.1$
Solar insolation	10–400 BTU/hr-ft²	Reference standard pyrheliometer	$5 \times 10^2 = 0.2$
Collector area	1–10^6 ft²	Linear rule	$1.7 \times 10^{-7} = 6 \times 10^{-6}$

Because most solar collector test facilities will not have available the equipment necessary to measure with these limiting accuracies, it is necessary to examine the accuracy and precision of measuring techniques which are available and in common use.

TEMPERATURE

INSTRUMENTAL ACCURACY AND PRECISION

Table C–2 lists the accuracy and precision required for temperature measurement according to ASHRAE 93-77 standards. The instrument accuracy represents the ability of the instrument to indicate the true value of the temper-

TABLE C–2 Temperature Measurement Requirements

	Instrument Accuracy	Instrument Precision
Temperature	$\pm 0.5°C$ ($\pm 0.9°F$)	$\pm 0.2°C$ ($\pm 0.36°F$)
Temperature difference	$\pm 0.1°C$ ($\pm 0.18°F$)	$\pm 0.1°C$ ($\pm 0.18°F$)

ature. The instrument precision represents the required closeness of agreement among repeated measurements of the same temperature.

When instruments, signal conditioners, and readouts are sequentially combined, their combined error is calculated on both an absolute sum basis (for non-random error) and on an RMS basis (random error assumption). In addition to errors introduced by instrumentation, there are errors resulting from insufficient response time or the coupling of the temperature measuring element to the object in which the temperature is being measured.

THERMOWELLS

The accuracy of temperature measurement varies with response time in a manner that differs for each method employed. For those methods that are most commonly considered the response time is proportional to the mass and specific heat of the temperature sensing element. If the thermometer is enclosed in a thermowell, the response time of the well is generally greater, i.e., it responds more slowly than that of the temperature sensing element within it, so the well determines the overall response time of the unit.

The use of thermowells for the measurement of fluid temperature is considered good practice for the following reasons.

1. The well provides a fixed geometry for fluid flow, no matter the type of temperature sensor placed within it. Also, sensors can be changed or moved from well to well without affecting fluid flow.

2. The well protects the temperature sensor from corrosive or chemically reactive components in the fluid which may affect the measurement or reduce the life of the sensor.

3. The well surface area provides an average over a certain region of fluid which will be representative of that region.

If fast response time is necessary, small temperature sensors can be placed directly in the fluid. Bare electrical leads in electrically conductive fluids, however, should be avoided.

Temperature sensors with a 1/e response time of 30 seconds are readily available. With this response time, the sensor will represent the fluid temperature to within 0.1°F after a 20°F step change in a time of 2 minutes and 40 seconds. The first order response time of a flat plate solar collector is generally proportional to flow rate, but it will usually fall in the 1 to 10 minute range.

There is a potential source of error through thermowell conduction and convection, but the error is proportional to the temperature difference between the fluid and the tubing wall so it diminishes with insulation of the tubing. Good practice leading to higher accuracy also requires thorough fluid mixing within the tubing. Two right-angle bends or staggered multiorifice plates can be used to achieve the required fluid mixing.

LIQUID MASS FLOW

BUCKET AND STOPWATCH

This technique of measuring mass flow is potentially very accurate since it uses direct measures of weight and time, each of which can be measured to reasonable accuracy. The accuracy of a manually operated stopwatch, for example, is limited primarily by variables in human response but can be regarded as within ±0.2 second with the percentage accuracy dependent upon the time period used to collect the fluid in the bucket. In practice, accuracy of 0.1% is achieved if a fluid mass on the order of 50 pounds is weighed. This method provides a time-averaged flow rate and is used to calibrate other flow measuring instruments in conjunction with constant head sources. An accuracy of 0.1% using a stopwatch would require the use of an estimated minimum of 5-minute time intervals.

HEAD FLOWMETERS

The output reading of a head flowmeter cannot be considered as uniformly accurate since the reading is affected by small changes in the geometry of the orifice plate, nozzle, or surrounding manifolds. Errors are also magnified by the effects of wear, corrosion, deposition, and stress on the components of a head flowmeter, in addition to the instrument's sensitivity to numerous fluid parameters, such as viscosity mixing ratios, Reynolds number, and temperature. At the low flow rates characteristic of solar collectors, head flowmeters become less accurate and impose greater restrictions on flow. The instrument's time response to step changes in flow can be less than one second.

VARIABLE AREA (ROTAMETER) FLOWMETER

The measurement of liquid flow with a rotameter is sensitive to many of the same factors as those affecting a head flowmeter while the direct viewing characteristic and the larger ratio of orifice perimeter to orifice area reduces the sensitivity of the instrument to changes at points along the orifice perimeter. The accuracy of a rotameter is properly expressed as a measure of the scale length so that the percentage of accuracy varies with the reading; the greatest accuracy is achieved for readings near the top of its scale. Also, the time response to a step change in flow is not the same in both directions for a rotameter. It is faster for a step increase than for a step decrease because the response time for a step decrease in flow is determined by the gravitation restoring force response time, whereas the response time for a step increase is determined by fluid mechanical response time. This will introduce a systematic error for integrated flow measurement.

MAGNETIC FLOWMETERS

These flowmeters create no obstruction in the flow path, will measure flow in either direction, and can be used with difficult liquids. But they will operate only in an electrically conducting fluid and their accuracy is limited by the stability of the fluid's electrical conductivity.

TURBINE FLOWMETER

Turbine flowmeters are potentially accurate, assuming proper maintenance addressed to bearing friction and changes in the rotor due to deposits and corrosion, with good experience obtained at low flow rates of non-aqueous, self-lubricating liquids. The time interval for sampling pulses must be long to achieve high accuracy because the electrical pulses are most often discretely counted. Accuracy is thus restricted to ± 1 count with percentage error from this source decreasing with an increasing time to accumulate counts.

ACOUSTIC FLOWMETERS

The accuracy of these flowmeters is limited by the accuracy to which the acoustical properties of the liquid are known.

ANGULAR MOMENTUM FLOWMETERS

Flowmeters using various angular momentum principles are available, offering a direct measure of mass flow. Correction factors are reduced and accuracies of better than 1% are possible with these instruments, according to some manufacturers' claims. Special designs using unobstructed flow paths are especially suited for biomedical instrumentation where low flow rate measurements are common.

RADIOISOTOPE FLOWMETERS

These flowmeters, used in conjunction with dye injection techniques, are capable of accuracies in proportion to the fluid path length available; the limit on accuracy is imposed by the dispersion and diffusion of the injected material in the fluid. Note that dispersion and diffusion rates are sensitive to many liquid properties which are themselves subject to significant error.

THERMAL FLOWMETERS

The thermal properties of liquids are somewhat more stable and better known than are acoustic, magnetic, or other properties used in flowmeter technology. Thermal flowmeters, therefore, are sometimes regarded as more accurate by comparison, but the improvement in accuracy is not a significant one.

POSITIVE DISPLACEMENT FLOWMETERS

The principal error in any positive displacement flowmeter is due to leakage around the moving member that is displaced during liquid flow, but there are other sources of error associated with the conversion of volume flow to mass flow, especially if air is mixed with the liquid. The time response of a positive displacement flowmeter is nearly instantaneous to step change in liquid flow, but these meters do not measure the actual rate of flow; they measure the amount of flow past a preselected time base. Thus, they are most often used to measure an accumulation of flow volume over a period of time. A true rate meter would require a differentiation of the shaft output of a positive displacement flowmeter so the instantaneous accuracy of these meters is limited by the accuracy of the differentiation. The response time for a positive displacement flowmeter used to measure the amount of flow is essentially determined by the time period used.

The data relative to all these various measurements of liquid volume or liquid mass flow, as well as additional information on pressure drop across the instrument and required corrections, are contained in Table C–3.

TABLE C–3 Accuracy and Precision of Liquid Flow Techniques

Method	Range (lb/min)	Quantity Measurement	Pressure Drop (lb/in^2)	Time Response*	Corrections Required	Estimated Accuracy % of Reading	Estimated Accuracy % of Span	Estimated Precision % of Reading	Estimated Precision % of Span
Bucket & stopwatch	10^{-6}–10^3	mass/time	<2	>5 m	None	±0.2			±0.1
Head flowmeters	10–10^6	pressure diff.	<5	<1 s	5	±5			±0.5
Variable area	2×10^{-3}–10^3	pressure diff.	<3	<10 s	6		±2		±0.25
Magnetic	10^{-3}–10^6	velocity	0	<1 s	5		±5		±0.5
Turbine	10^{-1}–10^5	velocity	<1	>10 s	3		±0.1		±0.01
Acoustic	10–10^6	velocity	0	<1 s	5	±5		±.5	—
Angular momentum	10^{-2}–10^6	mass flow	<1	<1 s	2	±0.25			±0.01
Radioisotope	10^{-2}–10^6	velocity	0	<1 m	5	±5		±.5	—
Thermal	10^{-1}–10^6	heat loss	0	<1 m	6	±3		±.5	—
Positive displacement:									
nutating disk	16–1280	vol. flow	<3	<1 m	3	±3		±0.1	—
oscillating piston	6–120	vol. flow	<3	<1 m	3	±0.2		±.015	—
lobed impeller	64–1.8×10^5	vol. flow	<3	<1 m	3	±0.2		±.015	—

* Within range suitable for solar collectors.

SOLAR INSOLATION

As has been noted previously, the most accurate means of measuring solar insolation over a period of time is the use of a continuously operating solar radiometer. The accuracy and precision of solar radiometers are compared in Table C–4, which also includes the specifications of the World Meteorological Organization (WMO) for different classifications of these instruments.

Although the limiting accuracy in measurements of the solar constant is unknown, terrestrial measurement of solar energy need not distinguish between effects that are intrinsic or secondary, as long as the energy received is locally measurable. To that end, the reference standard pyrheliometer is the most accurate instrument classification and is regarded as the standard upon which the accuracy of the others is based. The accuracy of this standard is based on calculations from first principles rather than actual calibration to the solar constant. A comparison of absolute cavity solar pyrheliometers conducted by DSET, Inc., at its test site in the Great Sonoran Desert in November of 1978 provided a measure of variation. With three different types of instruments (14 instruments in total) from ten organizations, the variation between instruments was ±0.3%.

COLLECTOR AREA MEASUREMENT

Although the limiting accuracy and precision of area measurements are very high, it is important to consider these simple procedures very briefly.

TABLE C–4 Classification of Accuracy of Solar Radiometers Established by World Meteorological Organization

	Sensitivity ($mW\ cm^{-2}$)	Stability (%)	Temperature (%)	Selectivity (%)	Linearity (%)	Time Constant (max)	Cosine Response (%)	Azimuth Response (%)	Galvanometer	Millimeter (%)	Chronometer
Reference standard pyrheliometer	±0.2	±0.2	±0.2	±1	±0.5	25 s	—	—	0.1 unit	0.1	0.1 s
Secondary instruments											
1st class pyrheliometer	±0.4	±1	±1	±1	±1	25 s	—	—	0.1 unit	0.2	0.3 s
2nd class pyrheliometer	±0.5	±2	±2	±2	±2	1 min	—	—	0.1 unit	±1.0	—
										Errors in recording apparatus	
1st class pyranometer	±0.1	±1	±1	±1	±1	25 s	±3	±3		±0.3	
2nd class pyranometer	±0.5	±2	±2	±2	±2	1 min	±5–7	±5–7		±1	
3rd class pyranometer	±1.0	±5	±5	±5	±3	4 min	±10	±10		±3	
										Errors due to wind	
										%	%
1st class net pyrradiometer	±0.1	±1	±1	±3	±1	½ min	±5	±5		±0.3	±3
2nd class net pyrradiometer	±0.3	±2	±2	±5	±2	1 min	±10	±10		±0.5	±5
3rd class net pyrradiometer	±0.5	±5	±5	±10	±3	2 min	±10	±10		±1	±10

STEEL TAPES

Length can be measured with steel tapes to an accuracy of one part in 10^3 or 0.1% in the range of 1 meter. This results in an accuracy of 0.2% for rectangular areas if consistent, non-random error is assumed.

MANUFACTURING TOLERANCES

In large solar installations where the total collector area is comprised of a number of individual, discrete panels, area can be measured by multiplying one collector panel area by the number of panels. With manufacturing tolerances of $\pm 1/16''$, the accuracy of an area measurement for one 2×8-foot collector is $\pm 0.326\%$. For N such collectors, the accuracy is:

$$\pm \frac{0.325}{\sqrt{N}} \%$$

(C–1)

This error results in accuracy of more than $\pm 0.1\%$ at $N = 10$ collectors. In order to maintain an accuracy of $\pm 0.1\%$ in collector area measurements, it is recommended that the area of $\frac{\sqrt{N}}{3}$ of the collectors in a large array be measured with steel tapes.

ACCURACY, TOLERANCE, PRECISION, AND SENSITIVITY

Accuracy refers to the degree to which a measurement reflects the actual value being measured and is usually expressed as a percentage of the absolute values of the measurement. In the measurement of temperature, however, accuracy in percentage of absolute temperature is often translated into units of temperature rather than a percentage. Thus, a thermometer accurate to $\pm 0.1°C$ at 100°C is accurate to $\pm 0.1/(273 + 100) \times 100 = \pm 0.0268\%$. This results from the fact that the temperature scale is more easily calibrated at 0° and 100°C than at its absolute zero value. For very high temperatures, accuracies return to percentages.

Tolerance refers to the amount within which the measured value is known to fall. It is properly expressed as an amount, not a percentage. Thus, strictly speaking, when a temperature is measured to an "accuracy" of $\pm 0.1°C$, it is more proper to refer to it as being measured within a tolerance of $\pm 0.1°C$.

The distinction between the accuracy and tolerance of a measurement is related to the source of error. If the source is due to uncontrollable factors in either the instrumentation or procedure, accuracy is reduced. If the source of error is a determined or controllable factor, then it contributes to a greater tolerance. When a manufacturer provides a platinum resistance thermometer with resistance specified to provide temperature measurement within a "tolerance" of $\pm 0.5°C$, the accuracy can be made greater than $\pm 0.5°C$ by proper calibration.

The sensitivity of an instrument is the change in output per unit change in input. The minimum sensitivity is the degree of precision. Precision is used to denote the degree to which repeated measurements can distinguish the effects of minor changes in the experimental variables.

In a measurement, N, consisting of component measurements $U_1, U_2, \ldots$ U_n, each contributing a known error, δU, and related to a known relationship, f:

$$N = f(U_1, U_2, \ldots, U_N) \qquad \text{(C-2)}$$

The absolute error, to first approximation, is given by:

$$\delta N = \left| \delta U_1 \frac{\partial f}{\partial U_1} \right| + \left| \delta U_2 \frac{\partial f}{\partial U_2} \right| + \cdots + \left| \delta U_N \frac{\partial f}{\partial U_N} \right| \qquad \text{(C-3)}$$

For a flat plate solar collector the absolute error is determined as follows:

$$\eta = \frac{\dot{m} \cdot \Delta T}{I \cdot A_c} \qquad\qquad C_p = 1$$

$$\frac{\partial \eta}{\partial \dot{m}} = \frac{\Delta T}{I \cdot A_c} \qquad\qquad \frac{\partial \eta}{\partial I} = -\frac{\dot{m} \cdot \Delta T}{A_c \cdot I^2}$$

$$\frac{\partial \eta}{\partial \Delta T} = \frac{\dot{m}}{I \cdot A_c} \qquad\qquad \frac{\partial \eta}{\partial A_c} = -\frac{\dot{m} \cdot \Delta T}{I \cdot A_c^2}$$

The absolute error is:

$$\delta \eta = \left| \delta \dot{m} \frac{\Delta T}{I \cdot A_c} \right| + \left| \delta \Delta T \frac{\dot{m}}{I \cdot A_c} \right|$$
$$+ \left| \delta I \frac{\dot{m} \cdot \Delta T}{A_c \cdot I} \right| + \left| \delta A_c \frac{\dot{m} \cdot \Delta T}{I \cdot A_c} \right| \qquad \text{(C-4)}$$

or:

$$\frac{\delta \eta}{\eta} = \frac{|\delta m|}{\dot{m}} + \frac{|\delta \Delta T|}{T} + \frac{|\delta I|}{I} + \frac{|\delta A_c|}{A_c} \qquad \text{(C-5)}$$

Thus, the percentage error of the efficiency is the sum of the percentage errors of mass flow, m, temperature difference, ΔT, solar insolation, I, and collector area, A_c.

The absolute error represents the greatest possible error that would occur if all the measurement errors contributed to efficiency error in the same direction. If the measurement errors can be assumed to be random in both magnitude and direction, then the resultant error in efficiency can be expressed as the probable error or statistical error:

$$\delta N = \sqrt{\left(\delta U_1 \frac{\partial f}{\partial U_1} \right)^2 + \left(\delta U_2 \frac{\partial f}{\partial U_2} \right)^2 + \cdots + \left(\delta U_N \frac{\partial f}{\partial U_N} \right)^2} \quad \text{(C-6)}$$

The assumption of randomness implies a prior knowledge of the statistical distribution of measured values compared to some actual value. Such prior knowledge then implies that errors are truly indicators of the extent to which the actual value may differ from the measured value. This, in turn, implies that repeated measurements can serve to reduce error. Systematic or non-random elements of error cannot be treated in this manner.

The use of probable error is justified in situations where repeated measurements under identical conditions are possible and where measurements made with a sequence of instrumentation can be compared using a calibrated source, so that statistical properties can be determined and randomness can be ascertained. Unfortunately, solar experimentation cannot be repeated under identical conditions and random errors cannot be easily distinguished from systematic ones, so that one cannot always be certain that the probable error is applicable. It is for this reason that both the probable and the absolute error are included in the analysis. It is fair to say that in most cases the probable error will represent the actual error to a sufficient degree and that the absolute error represents the limit of maximum error which may be encountered if unknown sources of systematic error are present.

In the example of the calculation of collector efficiency accuracy contained in Table C–5, in which several available instrumentation choices are shown, the combined errors of instrumentation used in sequence are determined by the probable error formula because calibrated sources can be used to verify their accuracy and precision. The final determination in efficiency measurement, however, requires both the probable and absolute error, since the error introduced by the sensors may be subject to systematic as well as random factors.

In summary, Table C–5 shows the accuracy of efficiency measurement for two sets of assumptions regarding measurement accuracies that may be considered typical in practice. The data show that error increases as ΔT becomes small, and that the errors will be significant if the error in ΔT is not kept well below 1°F, and if non-random errors are present. It becomes obvious that every effort should be made to provide an accurate measurement of ΔT to eliminate sources of systematic error and that flow rates should be low enough so that ΔT will be in a range where possibilities for error are reduced.

TABLE C–5 Collector Efficiency Accuracy (Example)

Solar insolation, $I = 300$ BTU/hr-ft², $\dfrac{\delta I}{I} = \pm 0.02$

Mass flow, $m = 15$ lb/hr-ft² $\times$ 16 ft² $= 240$ lb/hr, $\dfrac{\delta m}{m} = \pm 0.01$

Temperature difference, $\Delta T = T_0 - T_i = \dfrac{nIA_c}{m}$, $\dfrac{\delta \Delta T}{\Delta T} = \dfrac{1.0}{\Delta T}$ & $\dfrac{0.2}{\Delta T}$ (two cases)

Collector area, $A_c = 16$ ft², $\dfrac{\delta A_c}{A_c} = \pm 0.002$

Solar Collector Efficiency η	For Flow Rates Specified by ASHRAE 93-77 ΔT	Error in ΔT (Two cases) $\dfrac{\delta \Delta T}{\Delta T}$		Absolute (Maximum) Percent Error in Efficiency $\dfrac{\delta \eta}{\eta} \times 100$		Statistical (Random) Percent Error in Efficiency $\dfrac{\delta \eta}{\eta} \times 100$	
		$\delta \Delta T = 1.0°F$	$\delta \Delta T = 0.2°F$	$\delta \Delta T = 1.0°F$	$\delta \Delta T = 0.2°F$	$\delta \Delta T = 1.0°F$	$\delta \Delta T = 0.2°F$
0.7	14°F	0.071	0.014	±10.3	±4.6	±7.5	±2.7
0.6	12	0.083	0.017	±11.5	±4.9	±8.5	±2.8
0.5	10	0.100	0.020	±13.2	±5.2	±10.25	±3.0
0.4	8	0.125	0.025	±15.7	±5.7	±12.7	±3.4
0.3	6	0.167	0.033	±19.9	±6.5	±16.8	±4.0
0.2	4	0.250	0.500	±28.2	±8.2	±25.1	±5.5

APPENDIX D
PROPERTIES OF MATERIALS

TABLE D–1 Transparent or Translucent Cover Plate Materials

Properties	Ordinary Float Glass	White Low Iron Glass	Acrylic Plastic (Plexiglas)	Fiberglass: Reinforced Plastic	Polycarbon (Lexan)	Polyvinyl Fluoride (Tedlar)	Polyester Polyethylene (Mylar)	Fluorinated Hydrocarbons (Teflon)
Thickness	1/8″	1/8″	1/8′	25 mils	1/8″	4 mils	5 mils	1 mil
Solar transmission % at normal incidence	85/79	91/88	89/80	75–80	73	88	80	96
Transmission wavelengths (microns)	0.35–0.65/ 0.2–4.0	0.35–0.65/ 0.2–4.0	0.4–1.1/ 0.2–4.0	0.2–4.0	—	0.2–4.0	0.2–4.0	0.2–4.0
Infrared transmission (%)*	2	2	2	3–8	2	≅33	17.8	58
Infrared emissivity	0.88	0.88				0.59		0.35
Index of refraction (Air = 1.00)	−1.52	−1.51	1.49–1.56	1.54	1.59	1.46	1.64–1.67	1.3
Thermal conductivity (BTU in/hr-ft² °F)	65–100	65–100	1.3	NA	NA	NA	NA	1.35
Thermal expansion coefficient (in/in °F) × 10⁻⁶	4.8	5–8	19–46	14–20	37.5	28	9–15	59–90
Density (lb/in-ft²) Water = 5.2	13	13	6	7.5	6.2	7	7	11
Strength: Yield (psi)	1600/6400†	1600/6400	10,500	16,000	9,500	13,000	24,000	3,000
Elastic modulus (10⁶ psi)	10.5	10.5	0.45	1.1	0.35	0.26	0.55	0.07
Temperature limit (°F)	400	400	180	200	250	227	300	400
Approximate life (years)	500	500	10–15	7–20	10–15	5–10	4	15–20

* In wavelengths from 3–50 microns.
† Tempered.
NA: Not Available.

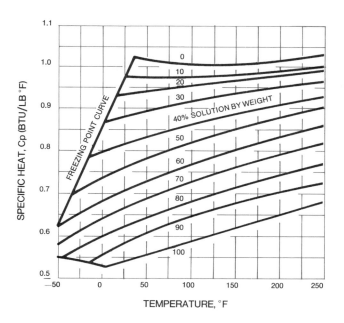

FIGURE D–1 Properties of ethylene glycol in water solution.

TABLE D–2 Absorber Plate Metals

	Copper	Aluminum	Steel	Iron	Black Rubber Plastic
Thermal conductivity at 212°F (BTU in/hr-ft² °F)	2616	1430	100–400	439	3–10
Thermal expansion coefficient ($°F^{-1} \times 10^{-6}$)	9.2	13.9	9.6	6.7	—
Specific heat, C (BTU/lb °F)	0.091	0.21	0.11	0.10	—
Density (lb/in-ft²)	46.5	14.1	40.7	40.9	4-7
Electrode potential in water (volts)	−0.340	+2.30	0	+0.441	0
Strength: Yield (psi) Elastic modulus (10^6 psi)	10–20,000 15.6	30,000 10	50–100,000 27–30	30–95,000 23–27	— —

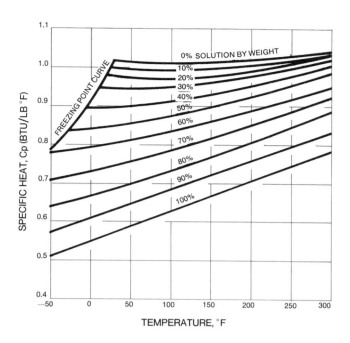

FIGURE D–2 Properties of propylene glycol in water solution.

TABLE D–3 Frame Materials

	Steel	Aluminum	Wood	Concrete
Strength: Yield (psi)	50–100,000	30,000	8–20,000	500–700
Elastic modulus (10^6 psi)	27–30	10	1.1–2.1	2.5–3.5
Thermal conductivity (BTU in/hr-ft^2 °F)	100–400	1430	0.7–2.4	6–8.5
Thermal expansion (in/in °F) × 10^{-6}	9.6	13.9	1.1–5.3	6
Density (lb/in-ft^2)	40.7	14.1	2.5–4.5	9–13

TABLE D–4 Densities and Thermal Conductivities

Thermal Conductivity (k) of Miscellaneous Substances at Room Temperature

Material	Density at 68°F (lb per cu ft)	Conductivity (BTU in/hr-ft²-°F)
Air, still	—	0.0169–0.215
Aluminum	168.0	1404–1439
Asbestos board with cement	123	2.7
Asbestos, wool	25.0	0.62
Brass, red	536.0	715.0
Brick		
Common	112.0	5.0
Face	125.0	9.2
Fire	115.0	6.96
Bronze	509.0	522.0
Cabots	3.4	0.25
Cellulose, dry	94.0	1.66
Celotex (sugar cane fiber)	13–14	0.34
Charcoal		
Coarse	13.2	0.36
6 mesh	15.2	0.37
20 mesh	19.2	0.39
Cinders	40–45	1.1
Clay		
Dry	63.0	3.5–4.0
Wet	110.0	4.5–9.5
Concrete		
Cinder	97.0	4.9
Stone	140.0	12.0
Corkboard	8.3	0.28
Cornstack, insulating board	15.0	0.24–0.33
Cotton	5.06	0.39
Foamglas	10.5	0.40
Glass wool	1.5	0.27
Glass		
Common thermometer	164.0	5.5
Flint	247.0	5.1
Pyrex	140.0	7.56
Gold	1205.0	2028.0
Granite	159.0	15.4
Gypsum, solid	78.0	3.0
Hair felt	13.0	0.26
Ice	57.5	15.6
Iron, cast	442.0	326.0
Kapok	1.0	0.24
Lead	710.0	240.0
Leather, sole	54.0	1.1
Lime		
Mortar	106.0	2.42
Slaked	81–87	—

TABLE D–4 Densities and Thermal Conductivities *(continued)*

Thermal Conductivity (k) of Miscellaneous Substances at Room Temperature

Material	Density at 68°F (lb per cu ft)	Conductivity (BTU in/hr-ft²-°F)
Limestone	132.0	10.8
Marble	162.0	20.6
Mineral Wool		
Board	15.0	0.33
Fill-type	9.4	0.27
Nickel	537.0	406.5
Paper	58.0	0.9
Paraffin	55.6	1.68
Plaster		
Cement	73.8	8.0
Gypsum	46.2	3.3
Redwood Bark	5.0	0.26
Rock Wool	10.0	0.27
Rubber, hard	74.3	11.0
Sand, dry	94.6	2.23
Sandstone	143.0	12.6
Sawdust	8–12	0.41
Sil-O-Cel (powered diatomaceous earth)	10.6	0.31
Silver	656.0	2905.0
Soil		
Crushed quartz (4% moisture)	100.0	11.5
Dakota sandy loam		
(4% moisture)	110.0	6.5
(10% moisture)	110.0	13.0
Fairbanks sand		
(4% moisture)	100.0	8.5
(10% moisture)	100.0	15.0
Healy clay		
(10% moisture)	90.0	5.5
(20% moisture)	100.0	10.0
Steel		
1% C	487.0	310.0
Stainless	515.0	200.0
Tar, bituminous	75.0	—
Water, fresh	62.4	4.1
Wood		
Balsa	7.3	0.33
Fir	34.0	0.8
Maple	44	1.2
Red Oak	48.0	1.1
White pine	32	0.78
Wood fiberboard	16.9	0.31
Wool	4.99	0.264

TABLE D–5 Solar Absorptivity and Thermal Emissivity of Commonly Used Metals

Metal Surface	Solar Absorptivity, α	Solar Reflectivity*	Thermal Emissivity ϵ
Aluminum, pure	0.1	0.9	0.1
Aluminum, anodized	0.12–0.16	0.84–0.88	0.65
Chromium	0.4	0.6	0.2
Copper, polished	0.15	0.85	0.03
Gold	0.2–0.23	0.77–0.8	0.025–0.04
Iron	0.44	0.56	0.07–1.1
Nickel	0.36–0.43	0.57–0.64	0.1
Silver, polished	0.035	0.965	0.02
Zinc	0.5	0.5	0.05

* Both specular and diffused reflectivity.

TABLE D–6 Thermal Storage Materials

Media	Temperature Range, °F	Melting Temperature °F	Latent Heat BTU/lb	Specific Heat C_p BTU/lb-°F	Density lb/ft³
Water ice	32	32	144	0.49	58
Water	90–130	32	—	1.0	62
Steel (scrap iron)	90–130	—	—	0.12	489
Basalt (lava rock)	90–130	—	—	0.20	184
Limestone	90–130	—	—	0.22	156
Paraffin wax	90–130	100	65	0.7	55
Salt hydrates					
$NaSO_4 \cdot 10H_2O$	90–130	90	108	0.4	90
$NA_2S_2O_3 \cdot 5H_2O$	90–130	120	90	0.4	104
$NA_2HPO_4 \cdot 12H_2O$	90–130	97	120	0.4	94
Water	110–300	—	—	1.0	62
Fire brick	110–800	—	—	0.22	198
Ceramic oxides M_6O	110–800	—	—	0.35	224
Fused salts $NaNO_3$	110–800	510	80	0.38	140
Lithium	110–800	370	286	±1.0	33
Carbon	110–800	6750	—	0.2	140
Lithium hydride	1260	1260	1200	±1.0	36
Sodium chloride	1480	1480	233	0.21	135
Silicon	2605	2605	607	—	146

NOTE: By mixing fused salts, any desired melting point between 250°F and 1000°F can be attained.

BIBLIOGRAPHY

The Widening Energy Gap and Alternative Energy Sources

Cheney, Eric S. "U.S. Energy Resources: Limits and Future Outlook." *American Scientist* 62 (January–February 1974): 14.

Hottel, H. C., and Howard, J. B. *New Energy Technology: Some Facts and Assessments.* Cambridge: The MIT Press, second printing 1972.

Reed, C. B. *Fuels, Minerals, and Human Survival.* Ann Arbor: Ann Arbor Science, 1975.

Rose, David J. "Energy Policy in the U.S." *Scientific American* 230, no. 1 (January 1974):20.

Starr, Chauncey. "Energy and Power." *Scientific American* 225, no. 3 (September 1971):37.

Thirring, Hans. *Energy for Man: From Windmills to Nuclear Power.* New York: Harper & Row, 1976.

The Prospects for Solar Energy

Bradford, P. V. "Investing in Solar Energy." Securities Research Division of Merrill Lynch, New York, NY, 1975.

Butt, Sheldon H. "Solar Heating and Cooling, A Developing Market." Presented before the Second Annual Meeting of the Solar Energy Industries Association in Washington, DC, 1975 Published by the SEIA.

Duffie, John A., and Beckman, William A. "Solar Heating and Cooling." *Science* 191, no. 4223 (16 January 1976):143.

Kaplan, Gadi. "Planning Solar's Future." *IEEE Spectrum* (June 1976):55.

Pollard, William G. "The Long Range Prospects for Solar Energy." *American Scientist* 64 (July–August 1976):424.

———. "The Long Range Prospects for Solar-Derived Fuels." *American Scientist* 64 (September–October 1976):509.

Stoll, Richard D. "Solar Collector Manufacturing Activity January Through June 1976." Nuclear Energy Analysis Division, Office of Coal, Nuclear & Electric Power Analysis, Federal Energy Administration, Room 207, Old Post Office Building, Washington, DC 20461.

The Sun

Flammarion, G. Camille, ed. *Flammarion Book of Astronomy.* New York: Simon & Schuster, 1964.

Lindsay, Sally, ed. "The Turbulent Sun." *Natural History* (November 1976): 54–81.

The Solar Spectrum and the Solar Constant

Coulson, Kinsell L. *Solar and Terrestrial Radiation.* New York: Academic Press, 1975.

Duffie, John A., and Beckman, William A. *Solar Energy Thermal Processes.* New York: John Wiley & Sons, 1974.

Kreider, Jan F., and Kreith, Frank. *Solar Heating and Cooling.* New York: McGraw Hill, 1975.

Blackbody Radiation Laws

Coulson, Kinsell L. *Solar and Terrestrial Radiation*. New York: Academic Press, 1975.
Duffie, John A., and Beckman, William A. *Solar Energy Thermal Processes*. New York: John Wiley & Sons, 1974.

Apparent Motion of the Sun

The American Ephemeris. Published each forthcoming year by the U.S. Government Printing Office, Washington, DC.
American Society of Heating, Refrigerating and Air Conditioning Engineers, Inc. *Handbook of Fundamentals*. 1972. 345 E. 47th Street, New York, NY 10017.
Brinkworth, B. J. *Solar Energy for Man*. Somerset, NJ: John Wiley & Sons, 1972.
Kreider, Jan F., and Kreith, Frank. *Solar Heating and Cooling*. New York: McGraw Hill, 1975 (pp. 45–56).

Estimation of Solar Insolation

American Society of Heating, Refrigerating, and Air Conditioning Engineers, Inc. *Handbook of Fundamentals*. 1972. 345 E. 47th Street, New York, NY 10017.
Yellott, John I., ed. "Solar Energy Utilization for Heating and Cooling." Chapter 59, 1974 edition, Applications Volume, of the ASHRAE Guide and Data Book series. The American Society of Heating, Refrigerating and Air Conditioning Engineers, Inc., 345 E. 47th Street, New York, NY 10017. Also reprinted as a National Science Foundation Report by the U.S. Government Printing Office, pamphlet NSF 74–41, stock number 038-000-00188-4.

Shadow Sizes

Waugh, Albert E. *Sundials, Their Theory and Construction*. New York: Dover Publications, 1973.

Measurements of Solar Insolation

Bennett, Iven. "Monthly Maps of Mean Daily Insolation for the United States." *Solar Energy* 9, no. 3 (July–September 1965):145.
Coulson, Kinsell L. *Solar and Terrestrial Radiation*. New York: Academic Press, 1975.
Klein, S. A. "Calculation of Flat Plate Utilizability." *Solar Energy* 21:393–402.
———. "Calculation of Monthly Average Insolation on Tilted Surfaces." *Solar Energy* 19:325–329.
Liu, B. Y. H., and Jordan, R. C. "Daily Insolation of Surfaces Tilted toward the Equator." *ASHRAE Transactions* 526–541 (1962).
———. "The Long Term Average Performance of Flat Plate Solar Energy Collectors." *Solar Energy* 7:53–74.
Lof, G. O. G., Duffie, J. A., and Smith, C. O. "World Distribution of Solar Radiation." July 1966. Report no. 21, Solar Energy Laboratory, University of Wisconsin, Engineering Experiment Station.
Quinlan, Frank. "Availability of Solar Radiation Data in the United States." Paper PH-79-8, No. 3, *ASHRAE Transactions*, Volume 85, Part 1.

Climatic Variables

Climatic Atlas of the United States. Reprinted by the National Oceanic and Atmospheric Administration, U.S. Department of Commerce. Available from the National Climatic Center, Federal Building, Asheville, NC 28801.
Engineering Design Manual for Outdoor Advertising Structures. Published by the Outdoor Advertising Association of America, Inc., 625 Madison Ave., New York, NY 10022. (See section on wind loads.)

Statistical Abstract of the United States. Published annually by the U.S Department of Commerce, Bureau of the Census. Available from the U.S. Government Printing Office, Section G, Geography and Environment.

Diffuse and Reflected Components of Solar Isolation

Applications Handbook. Published by the American Society of Heating, Refrigerating and Air Conditioning Engineers, Inc., 345 E. 47th Street, New York, NY 10017 (Chapter 59.)

Kreider, J. F., and Kreith, F. *Solar Heating and Cooling.* New York: McGraw Hill, 1975 (p. 77).

System Approximations

"Solar Energy for Space Heating and Hot Water." 1976. SE-101. Published by the Energy Research and Development Administration, Washington, DC.

Flat Plate Collectors

Hottel, H. C., and Woertz, B. B. "The Performance of Flat-Plate Solar Heat Collectors." *Transactions of the American Society of Mechanical Engineers* 64 (1942):91.

Leckie, Jim, et al., *Other Homes and Garbage.* San Francisco: Sierra Club Books, 1975.

Selective Surfaces

Baumeister, Philip. "A Comparison of Solar Photothermal Coatings." *Proceedings of the Society of Photo-Optical Instrumentation Engineers* 85 (1976):47–61.

McDonald, Glen E. "Selective Coating for Solar Panels." National Technical Information Service Report N76-15603, December 1975. Obtained from NTIS, 5285 Port Royal Road, Springfield, VA 12161.

Schreyer, J. M., et al. "Selective Absorptivity of Carbon Coatings." Report Y/DA-6701, July 1976. Oak Ridge Y-12 Plant, Oak Ridge, TN. Available from the Union Carbide Corporation Nuclear Division or from the Energy Research and Development Administration.

Tabor, H. "Selective Surfaces for Solar Collectors." Chapter IV of *Low Temperature Engineering Application of Solar Energy.* The American Society of Heating Refrigerating and Air Conditioning Engineers, Inc., 345 E. 97th Street, New York, NY 10017.

Concentrating Collectors

Nelson, D. T., Evans, D. L., and Bansal, R. K. "Linear Fresnel Lens Concentrators." *Solar Energy* 17, no. 5-C (1975):285.

Rabl, A. "Comparison of Solar Concentrators." *Solar Energy* 18 (1976):93.

Winston, R. "Light Collection Within the Framework of Geometrical Optics." *Journal of the Optical Society of America* 60 (1970):245.

Photovoltaic Collectors

Chalmers, Bruce. "The Photovoltaic Generation of Electricity." *Scientific American* 235, no. 4 (October 1976):34.

Measurement Techniques and Procedures

Considine, Douglas M. *Process Instruments and Controls Handbook.* New York: McGraw Hill, 1974.

Duebelin, Ernest O. *Measurement Systems—Applications and Design.* New York: McGraw Hill, 1975.

"Methods of Testing for Rating Solar Collectors Based on Thermal Performance." NBSIR 74-635, December 1974. Interim Report prepared for the National Science Foundation by the National Bureau of Standards.

"Method of Testing Solar Collectors Based on Thermal Performance." ASHRAE Proposed Standard 93 P, June 1, 1976.

General References

Agosto, William N. "Microwave Power: A Far-Out System." *IEEE Spectrum* (May 1976):48.

Agranoff, Joan, ed. *Modern Plastics Encyclopedia 1975–1976*. New York: McGraw Hill, 1976.

Balcomb, J. Douglas, and Hedstrom, James C. "Sizing Collectors for Space Heating." *Sunworld* no. 1 (July 1976):25. Published by the International Solar Energy Society, P.O. Box 26, Highett, Victoria 3190 Australia.

Baumeister, Theodore, ed. *Mark's Standard Handbook for Mechanical Engineers*. 7th ed. New York: McGraw Hill, 1967; 8th ed., 1978.

Considine, Douglas M. *Process Instruments and Controls Handbook*. New York: McGraw Hill, 1974.

Duffie, John, and Beckman, William. *Solar Energy Thermal Processes*. New York: John Wiley & Sons, 1974.

Gunther, Raymond C. *Refrigeration, Air Conditioning, and Cold Storage*. Chilton Books, 1969.

Hicks, Tyler G., ed. *Standard Handbook of Engineering Calculations*. New York: McGraw Hill, 1972.

Kays, W. M., and London, A. L. *Compact Heat Exchangers*. 2d ed. New York: McGraw Hill, 1964.

Klein, S. A., Beckman, W. A., and Duffie, J. A. "Design Procedure for Solar Heating Systems." *Solar Energy* 18, no. 2 1976:113.

Kreider, Jan F., and Kreith, Frank. *Solar Heating and Cooling*. New York: McGraw Hill, 1975.

Materials Engineering, Vol 81. Materials Selector, mid-September 1975. Published by Reinhold Publishing Company, 600 Summer Street, Stamford, CT 06904.

Meinel, Aden B., and Meinel, Marjorie P. *Applied Solar Energy*. Reading, Mass: Addison-Wesley, 1976.

Peny, Robert H., and Chilton, Cecil H. *Chemical Engineer's Handbook*. 5th ed. New York: McGraw Hill, 1973.

Ratzel, A. C., and Bannerot, R. B. "Optimal Material Selection for Flat Plate Solar Energy Collectors Utilizing Commercially Available Materials." Presented at the 16th National Heat Transfer Conference, St. Louis, August 8–11 1976. Available from the American Institute of Chemical Engineers, 345 E. 47th Street, New York, NY 10017.

Tabor, H. "Radiation, Convection, and Conduction Coefficients in Solar Collectors." *Bulletin of the Research Council of Israel* 6C:155, 1958.

Williams, J. Richard. *Solar Energy Technology and Applications*. Ann Arbor: Ann Arbor Science, 1977.

Weast, Robert C., ed. *Handbook of Chemistry and Physics*. CRC Press, 18901 Cranwood Parkway, Cleveland, Ohio 44128.

INDEX

Page numbers in **bold type** indicate information found in illustrations; page numbers followed by t indicate information found in tables.